MATTHEW H

Body Language

Your Great Guide For The World Of Body Language Psychology And The Different Techniques Of Dark Psychology and Non-Verbal Communication To Become The Master Of Your Success

Copyright © 2020 Matthew Hall

All rights reserved.

© **Copyright 2020 - All rights reserved.**

The content contained within this book may not be reproduced, duplicated or transmitted without direct written permission from the author or the publisher.

Under no circumstances will any blame or legal responsibility be held against the publisher, or author, for any damages, reparation, or monetary loss due to the information contained within this book. Either directly or indirectly.

Legal Notice:

This book is copyright protected. This book is only for personal use. You cannot amend, distribute, sell, use, quote or paraphrase any part, or the content within this book, without the consent of the author or publisher.

Disclaimer Notice:

Please note the information contained within this document is for educational and entertainment purposes only. All effort has been executed to present accurate, up to date, and reliable, complete information. No warranties of any kind are declared or implied. Readers acknowledge that the author is not engaging in the rendering of legal, financial, medical or professional advice. The content within this book has been derived from various sources. Please consult a licensed professional before attempting any techniques outlined in this book.

By reading this document, the reader agrees that under no circumstances is the author responsible for any losses, direct or indirect, which are incurred as a result of the use of information contained within this document, including, but not limited to, — errors, omissions, or inaccuracies.

Table of Content

Introduction 5

Chapter 1. How Non-Verbal Communication Works 13

Chapter 2. How to Understand People Through Body Language 23

Chapter 3. Manipulation Through Body Language 32

Chapter 4. Uses of Body Language 38

Chapter 5. Guide to an Effective Body Language 45

Chapter 6. How to Persuade People 52

Chapter 7. How to Analyze People 59

Chapter 8. Dark Psychology Secrets 66

Chapter 9. How to Defend Ourselves of Dark Psychology 73

Chapter 10. The Power of Hands 81

Chapter 11. Body Language of a Child 86

Chapter 12. Reading People Through Their Mind 92

Chapter 13. Myths About Body Language 100

Chapter 14. Who Is a Manipulator? 107

Chapter 15. Dark Psychology and Women 114

Chapter 16. Techniques of Dark Psychology 121

Chapter 17. Practical Exercises For Mind Control Prevention 128

Chapter 18. Touch As a Form of Body Language 134

Chapter 19. Facial Expression As a Form of Body 139

Chapter 20. Manipulation In a Relationship 147

Chapter 21. How to Know if Someone Is Lying Through Body Language 154

Chapter 22. How Body Language Improves Your Mindset 161

Chapter 23. Proxemics 169

Chapter 24. Muscular Core, Posture, and Breathing 172

Chapter 25. Hand Gestures and Arm Signals 177

Chapter 26. Different Types of People and How They Fit In the Social Circle 183

Conclusion 190

Introduction

Body language involves using our physical behavior, expressions, and manners to reveal nonverbal information about ourselves, which is usually done unconsciously. Many people are not mindful of it, but you are continually giving out body cues and wordless signals that reinforce the interaction or contradict what you are trying to say in all your interactions.

Your entire nonverbal behavior transmits a loud and strong message that continues even after you stop talking. There are instances when what someone says might differ from what their body language is communicating. Hence, in this case, it will be easy for the person you are interacting with to pass you off as a liar. If someone asked for a favor and you gave a smile after giving a no, you have ended up confusing the person. With this kind of mixed-signal, the person might be confused about what to believe. However, if the person understands the concept of body language, they would probably just walk away since the body language is unconscious and gives someone away by revealing their real intention.

The Essence of Nonverbal Communication

The cues you are unconsciously giving out from your body are pretty essential and, as said earlier, give meaning to the interaction you are in. From your body cues, the person you are with will know whether you are interested in the relationship or not, whether you are hiding information or being explicit, and whether you are paying attention.

With nonverbal communication signals that complement what you say, you can build trust, rapport, and clarity. I am pretty sure you know what happens when your words and body language cues contrast!

In reading body language signals, you have to notice the body language people are giving out. It does not stop at that, and you have to be sensitive to yours as well. In understanding nonverbal communication, pay attention to the following roles it plays:

Repetition

In other words, it enforces the message you are trying to pass. You made a marriage proposal to your girlfriend, for instance. After popping the question and she accepted, you would generally expect her to smile, jump up and be excited. However, if she said yes with a straight face, I am pretty sure you know something is not right.

Contradiction

It can also refute the message you pass across, thus giving the signal that you might be lying. You came home from a two-week journey. Your wife greeted you and said she was excited to see you, but without a hug, a smile, or any facial expression to corroborate the statement. Something was off.

Substitution

Body cues often stand in place of verbal communication. In African culture, for instance, let us assume a guest visited a family. As this person was leaving, he offered the child some money. The mother gave the child "that kind of look," and the child took it as a cue to reject the offer.

Complementing

Body language cues might add more weight to the meaning of the message you are passing across verbally. Consider a man who tells his wife, "I love you" and drives off. Another man plants a kiss on the wife's forehead and says, "I love you" while looking into her eyes. Of these two, it is clear which one meant what he was saying.

Accenting

Your nonverbal cues can also emphasize the point you are trying to make. Saying no, alongside a shaking of the head, emphasizes the weight of the negation.

Without beating around the bush any more, let us examine how you can read various clues from body parts.

Reading Various Parts of the Body

Head Movement

Head movement is one of the meekest body languages to decode. However, for someone who has no clue what this nonverbal communication signal means, hardly will they make sense of it. To explain the head movement, I have here two scenarios:

As part of the exercise to get a job, a candidate must decide why he is the best candidate for the job. During the presentation, his audience, the hiring manager, nods quickly while the candidate desperately keeps trying to sell himself. He is unaware of the hiring manager's message, which clearly shows he is wasting his time.

Consider another candidate giving the same presentation. As he goes off trying to sell himself, the hiring manager leans back with his head tilted. Oblivious to the meaning of this body language, the candidate does not try to shed light on the point that triggered the manager's body reaction. He is ignorant of the body language; hence, he keeps on blabbing.

Reading the Face

There are many expressions we can reveal with our faces. Even babies and toddlers are smart enough to decode this body language cue. That a smile reveals happiness or satisfaction or a frown shows dissatisfaction or sadness. There are times when the facial expression could give insight into what is going on with a person. A person who says they're fine with a slight frown, for instance, could be lying.

It is a universal expression that conveys a wide range of emotions, such as sadness, fear, panic, anxiety, worry, disgust, distrust, happiness, and many others. The best part is that this expression does not change or vary with people.

Many people, in a bid to hide their real intention, desperately try to control the face. However, a careful study of the face can give you a clear glimpse into the message someone is trying to pass across. There are times when someone might hide primary body language, such as raised eyebrows, smiles, frowns, etc. Be sure to look out for the subsequent:

A warm and genuine smile does light up the whole face. It indicates happiness. It is also an unmistakable symbol that the other party is enjoying your company.

On the other hand, a phony smile is a polite way of showing approval, even if they do not enjoy the conversation or interaction. For you to detect a phony smile, take a look at the side of the eyes. The lack of crinkles is all you need to pass a smile off as fake.

The Eye Window

The eyes disclose a lot regarding a person. It explains why the eyes are referred to as the window to the soul. Besides, it is an essential and natural communication process for you to note all interactions' eyes.

In communicating with people, it is customary to note eye contact, whether someone is averting your gaze or not, the rate of blinking, and their pupils' size.

The following explains some nonverbal clues from the eyes:

Eye gaze

A person interested and paying attention to a conversation will look directly into your eyes while having a conversation. However, they might break eye contact once in a while because prolonged eye contact is rather uncomfortable. A distracted and uninterested person, on the other hand, will often break eye contact and look away. This person might be awkward or is trying to hide their true feelings.

Blinking

While blinking is an entirely natural process, the frequency matters. A person uncomfortable or in distress will blink more often. On the other hand, infrequent blinking means that a person is intentionally trying to control their eye movements.

Pupil Size

Pay attention to pupil size as it is very subtle and affected by the room's light level. However, emotions also affect pupil dilation, causing small changes in the pupil's size. It explains why someone with highly dilated eyes is either aroused or interested in a person.

Hand Movements

Some cues can easily be found from the hand's position and pattern of movement. We explain this in detail:

When someone has their hands in their pockets, they could be lacking confidence, hiding information, or just being defensive.

When a person unconsciously points to another person in a group or meeting while making a speech, there might be some common ground they share.

In communicating with someone, there is the presence of an obstacle. It is in the form of an object between you, and the person translates to the person trying to block you out. In this circumstance, your goal should be to build rapport and gain such a person's trust.

A person talking with the palms facing up is likely, to be honest. Such a person is not hiding the palm since they most likely have nothing to hide.

The Mouth

The expressions and movement of the mouth are pretty vital in decoding body language as well. It is why a worried, anxious, or insecure person will likely chew their lower lip. Some forms of nonverbal communication cues from the mouth will be examined below. A person, to be polite, might cover the mouth if the other party is yawning. Be watchful, as it can be done to cover up a frown as well.

Pursed lips

When a person tightens up their lip, it could signal objection, disapproval, or distaste.

Lip biting

It is common when a person is anxious, worried, or stressed.

Covering the Mouth

It could be done to hide emotional reactions like smiles or smirks.

A Slight Change in Direction

A person's feelings can be seen through the direction of the mouth. As a result, someone happy or in a good mood might have their mouth slightly turned up. A slightly turned-down mouth, on the other hand, could signal sadness or displeasure.

The Importance of Reading People

The world is made of people. Life is better enjoyed when you have people to relate with. However, your survival in the world also depends on your ability to decide when not cooperating with some people. So, your ability to read people is essential.

There are times you are unconsciously cooperating with others. The fact that you walk gently to your place of work without causing a scene or doing anything to warrant unnecessary attention is an act of cooperation with the rest of the society on some level. You don't just awaken up one day and choose to go on a killing spree. You are connected to the Internet and the rest of the world alike. All these things require some form of human cooperation.

For this to occur, people unconsciously have to come to an appropriate form of agreement and acceptable behavior on some level. All in all, cooperating with people is pretty important, and your decision whether to cooperate or not comes down to your ability to read people.

The best salesman knows how to coax you because they are good at analyzing people. They can get you into buying what they have to offer, even if you do not need what they are offering. The better you are at reading other people's motives, the better you can deal with such a person.

Chapter 1. How Non-Verbal Communication Works

Being able to connect well is extremely important when wanting to succeed in the personal and professional world, but it isn't the words you say that scream. It is your body language that does the screaming. Your gestures, posture, eye contact, facial expressions, and tone of voice are your best communication tools. These can confuse, undermine, offend, build trust, draw others in, or put someone at ease.

There are many times where what someone says and what their body language says is different. Non-verbal communication could do five things:

- Substitute—It could be used in place of a verbal message.

- Accent—It could underline or accent your verbal message.

- Complement—It could complement or add to what you are saying verbally.

- Repeat—It could strengthen and repeat your verbal message.

- Contradict—It could go against what you are trying to say verbally to make your listener think you are lying.

Many different forms of Non-verbal communication will be looked at, and we are going to cover:

- Gestures—These have been woven into our lives. You might speak animatedly; argue with your hands, point, wave, or beckon. Gestures do change according to cultures.

- Facial expressions—You will learn that the face is expressive and shows several emotions without speaking one word. Unlike what you say and other types of body language, facial expressions are usually universal.

- Eye contact—Because sight tends to be our strongest sense for most people, it is an essential part of Non-verbal communication. The way someone looks at you could tell you whether they are attracted to you, affectionate, hostile, or interested. It might also help the conversation flow.

- Body movement and posture—Take a moment to think about how you view people based on how they hold their heads, stand, walk around, and sit. The way a person carries gives you much information.

Lower Body

The arms share much information. The hands share a lot more, but legs give us the exclamation point and tell us precisely what someone is thinking. The legs could say to you if a person is open and comfortable. They could also imply dominance or where they want to go.

Upper Body

Upper body language can show signs of defensiveness since the arms could easily be used as a shield. Upper body language could involve the chest. Let's look at some upper body language.

Leaning

If someone leans forward, it will move them closer to another person. There are two possible meanings to this. First, it will tell you that they are interested in something, which could just be what you are talking about. But this movement could also show romantic interest. Second, leaning forward could invade a person's personal space; hence, leading to a threat. It is often an aggressive display. It is done unconsciously by influential people.

The Superman

Bodybuilders, models commonly use this, and it was made famous by Superman. It could have various meanings depending on how a person uses it. Within the animal world, animals will try to make themselves look bigger when they feel threatened. If you look at a house cat when they get spooked, they will stretch their legs, and their fur stands on end. Humans also have this, even if it isn't as noticeable. It is why we get goosebumps. Because we can't make ourselves look bigger, we have to develop arm gestures like putting our hands on our waist. It shows us that a person is getting ready to act assertively.

The Chest in Profile

If a person stands sideways or at a 45-degree angle, they are trying to accentuate their chest. They might also thrust out their chest, more on this in a minute. Women do this posture to show off their breasts, and men will show off their profile.

Outward Thrust Chest

If someone pushes their chest out, they try to draw attention to this part of their body. It could also be used as a dreamy display. Women understand that men have been programmed to be aroused by breasts. If you see a woman pushing her chest out, she might be inviting intimate relations. Men will thrust out their chests to show off their chest and possibly trying to hide their gut. The difference is that men will do this to women and other men.

Hands

Human hands have 27 bones, and they are a very expressive part of the body. It gives us much capability to handle our environment.

Reading palms isn't about just looking at the lines on the hands. After a person's face, the hands are the best source for body language. Hand gestures are different across cultures, and one hand gesture might be innocent in one country but very offensive in another.

Hand signals may be small, but they show what our subconscious is thinking. A gesture might be exaggerated and done using both hands to illustrate a point

Face

People's facial expressions could help us figure out if we trust or believe what they are saying. The most trustworthy expression will have a slight smile and a raised eyebrow. This expression will sow friendliness and confidence.

We make judgments about how intelligent somebody is by their facial expressions. People who have narrow faces with a prominent nose were thought to be extremely intelligent. People who smile and have joyous expressions could be thought of as being smart rather than someone who looks angry.

Mouth

Mouth movements and expressions are needed when trying to read body language. Chewing on their lower lip might indicate a person who is feeling fearful, insecure, or worrying.

If they cover their mouth, this might show that they are trying to be polite if they are yawning or coughing. It might be an effort to cover up disapproval. Smiling is the best signal, but smiles can be interpreted in many ways. Smiles can be genuine, or they might be used to show cynicism, sarcasm, or false happiness.

Negative Emotions

The silent signals that you show might harm your business without you even knowing it. We have over 250,000 facial signals and 700,000 body signals. Having poor body language could damage your relationships by sending other signs that you can't be trusted. They might turn off, alienate, or offend other people.

You have to keep your body language in check, and this takes much effort. Most of the time, you may not know that you are doing it, and you might be hurting your business and yourself.

Here are some emotions and how to spot them:

Fear, Anxiety, or Nervousness

Fear could happen when our basic needs get threatened. There are many different levels of fear. Suppose might be mild anxiety or full-blown blind terror. The various bodily changes that get created by fear can make this one easy to spot.

- Voice trembling.

- Errors in speech.

- Pulse rate extremely high.
- Vocal tone variations.
- Sweating.
- Lips trembling.
- Muscle tensions like their legs wrapped around something, clenched hands or arms, elbows are drawn in, jerky movements.
- Damp eyes.
- Holding their breath or gasping for breath.
- Not looking at one another.
- Fidgeting.
- Dry mouth indicated by licking their lips, rubbing their throat, or drinking water.
- Defensive body language.
- Face is pale.
- Fight or flight body language.
- Breaking out in a cold sweat.
- Any symptoms of stress.

Sadness

- Lips trembling.

- The flat tone of voice.
- Body drooping.
- Tears.

Anger

- Clenched fists.
- Invading body space.
- Leaning forward.
- Baring their teeth or snaring.
- Using aggressive body language.
- Neck or face is red and flushed.
- Displaying power body language.

Embarrassment

- Not making eye contact.
- Looking down and away.
- Neck or face is red and flushed.
- Changing the subject or trying to hide their embarrassment.
- Grimacing.
- Fake smiles.

Positive Emotions

When you have positive body language, it means that you are engaging, approachable, and open. It isn't saying that you need to use this kind of body language all the time or that it is the best signs that will show a person is friendly. It's just a good beginning point for reading positivity in others as well as yourself.

Non-verbal Signals Used Universally

Non-verbal communication is different for everybody and in different cultures. A person's cultural background will define their non-verbal communication since some communication types, like signs and signals, need to be learned.

Since there are different meanings in non-verbal communication, there could be miscommunication when people from different cultures try to communicate. People might offend others without really meaning to due to cultural differences. Facial expressions are very similar around the world.

Seven micro expressions are universal, and we will go more in-depth about these, but they are hate/contempt, anger, disgust, surprise, fear, happiness, and sadness. It might be different in the extent of how people show these feelings since, in specific cultures, people might readily show them where others won't.

Nods might also have different meanings, and this can cause problems, too. In some cultures, their people might not say "yes," but people from different cultures will interpret as "no." If you nod in Japan, they will solve it as you are listening to them.

Here are other non-verbal communications and how they differ in various cultures:

Eye Contact

Many Western cultures consider eye contact as a good gesture. It shows confidence, attentiveness, and honesty. Cultures such as Hispanic, Asian, Native American, and Middle Eastern don't think eye contact is a good gesture. They believe it is rude and offensive.

Unlike Western cultures that think it's respectful, others don't think this way. In Eastern countries, women absolutely can't make eye contact with men since it shows the power or sexual interest. Many cultures accept gazes as only showing an expression, but staring is thought of as rude in most.

Gestures

You need to be careful doing a "thumbs up" because it is very different in many cultures. Some view it as meaning "okay," but in Latin America, it is vulgar. Japan views it as meaning money.

Snapping your finger may be acceptable in some cultures, but it is disrespectful and offensive in others. In some Middle Eastern countries, showing your feet can be offensive. Pointing your finger is an insult in some cultures. People in Polynesia will stick their tongue out when they greet someone, but other cultures see it as a sign of mockery.

Touch

Touch is thought of as rude in most cultures. Some cultures look at shaking hands to be acceptable. Kissing and hugs, along with other touches, are looked at differently in different cultures. Asians are too conservative with these types of communications.

Patting someone's head or shoulder has different meanings in different cultures. Patting a child's head in Asia is extremely bad since their head is the sacred part of their body. Middle Eastern countries think people of opposite genders touching to be horrible character traits.

How and where a person gets touched could change the meaning of that touch. You need to be careful if you travel to various places.

Appearance

It is an acceptable form of non-verbal communication. Their appearance has always judged people. Differences in clothing and racial differences could tell a lot about anyone.

Making yourself look good is an important personality trait in many cultures. What is thought to be a good appearance will vary from country to country. How modest you get is measured by your appearance.

Chapter 2. How to Understand People Through Body Language

The Capital Importance of Body Language

Our bodies cannot show anything but what is in us. Our emotions use the sounding board. Therefore, we understand that our body's non-verbal part of communication always reflects our mental state, whatever the situation. Indeed, where does it come from, if not our psyche?

Body language is subject to physical law: energy does not vanish; it transforms. As electricity becomes light, heat, or movement, our psyche becomes body language.

Beware, the same gesture may have different meanings. For example, a person with arms crossed, a gesture generally interpreted as a negative signal. Indeed, arms crossed the pass to outsource refusal, withdrawal, skepticism, antipathy, etc. Sometimes this interpretation is accurate, but not always. What about a man struck, waiting for the bus? Is it expressly rejected? If so, to whom? Facing the bus? To other people like him at the bus stop? What if nobody exists? In this situation, arms crossed indicate nothing but being able to do nothing but wait. No reason to move, our man folds his arms.

To correctly decipher body language, you must first consider the action context. Also, as with verbal language, an "expression" that does not fit into a situation will have a high potential for misunderstanding.

Incorrect posture can reveal insecurity, fear, distrust, etc. On the other hand, the right posture gives the impression of strength, power, and confidence.

Understand more about using body signals to convey the desired impression.

Negative Body Language

Often, during a conversation, you can pass negative body language without realizing it.

Facial expressions and gestures end up showing several details.

Some negative postures that you should avoid in your client meetings:

- Hands-on-hips or pockets;
- Knees pointing to the exit door;
- Legs wide open;
- Crossed arms.

These attitudes are perceived, even if unconsciously, by the other person and can ruin a sale's progress.

Know other signs you should avoid avoiding transmitting the wrong body language.

Hand to Mouth

Experts analyze that when a person is not telling the truth, they usually cover their mouths.

Disparities of this posture habit are:

- Rub your lips;
- Dash the chin;
- Put stuff in front of the mouth.

Compressed Lips

Another negative body language is to compress your lips.

This act shows that the person is trying to avoid saying what he thinks.

That is, hiding your lips reveals that you don't want to answer any questions.

Defocused Look

Body language says a lot by looking.

A look without focus, or looking up and to the right, indicates confusion.

It is because, when looking away, the person is looking for a mental image.

Therefore, he shows a lack of clarity in his speech, as well as insecurity.

Forehead Contracted

In conversation, if the other person wrinkles his forehead, that's not a good sign.

These horizontal lines show a certain level of tension, doubt, or nervousness, which is a bad sign of body language.

Restricted Hand and Arm Movements

Keeping your hands behind your back or clinging to your body conveys the message of little confidence.

Another gesture to be avoided is to put your hands back or feet crossed behind the chair.

These are signs of discomfort.

Reading Body Language

Reading body language may help to assess the feeling behind or instead of the words spoken. The adult can quickly and instinctively understand that a child is frightened by thunder when they see the child screaming and covering their ears. However, there are misconceptions about body language, causing miscommunication unless the whole-body language is read.

The eyes have long been named the windows to the soul, and it may be this concept that created the greatest myth in reading bodies. It is widely assumed that if a person avoids eye contact or does not hold it, that person does not say the facts. It is a mistaken assumption though popularly known. Pathological liars may maintain sustained contact with the eyes because they realize most people assume that looking away from the eyes shows an untruth. People who say the truth do not keep eye contact because they clearly state evidence and feel no need to convince anyone of this.

When a person is depressed or uncomfortable, and avoidance of eye contact occurs. For example, a child being chastised by a parent would always look down on the ground instead of him looking in the parent's eyes. Painfully shy or nervous people, too, are having a tough time meeting another person's eyes in conversation. Someone with little actual knowledge of reading body language would find the people lying by the standard error in each of these instances. Instead, the child shows authority, the adult, contrite, and the shy person shows typical distress or uncomforted sign.

In addition to the eye movement, the overall body language used needs to be looked at. Fidgeting, drumming fingers, or playing with hair matched to a lack of eye contact shows that a person is dissatisfied with the situation or discussion topic. Still, eyes fixed on a distant point by a person with arms crossed, attentive to the conversation, shows, instead, serious attention and intense thinking on the topic of discussion.

The precise interpretation of body language may provide an insight into another person's thoughts, feelings, and emotions. However, to read correctly, it is essential to remember to look at the entire body's movements, or words, rather than treating part of the body as separate from the other parts.

Keys to Reading Body Language

Body language is the type of communication a person uses to respond to circumstances, including facial expressions. More than 54% of the way we communicate with each other consists of body language, 39% consists of how the voice is used, and just 7% consists of the words spoken. Developing one's ability to interpret and understand gestures and signs of body language would greatly benefit because it will help better understand and interact with other human beings.

Body language involves body expressions, gestures, eye contact, muscle tension, skin coloring, breathing rate, etc. Of course, you should remember that body language is different from people to people and various nationalities and cultures. Consequently, it is at all times good to check what is seen in a person. It can be achieved by answering similar questions and endeavoring to better-known individuals.

There is also a lot of myths about interpreting body language. Most deceptive books and internet guides don't teach people the right thing. There's the truth you need to learn about this, while popular reasoning can trick you into believing a person's ability to read body language is the real secret to finding lies. Some important body language secrets are:

Posture

In most situations, if you take the correct pose, you should build the right impression on people. Leaning a little towards a person can create an image of friendliness. It could also be that you have an interest in others. At the same time, seeing a level head establishes a sense of self-assurance and trust.

Legs

When a person is anxious, the legs are always moving around. It happens when the person tells lies and is bored too. It is safer to keep those legs crossed or even to give the opposite impression, to appear confident and polite to others.

Eye contact

If you keep healthy eye contact, you show concern and respect for others. And you still need to find a balance. When you hold too much eye contact, the other person would feel self-conscious. Sometimes, if you don't have enough eye contact, you can make the other person believe you're not involved in what's being addressed.

Arms

If you have primary arms crossed, you can make yourself look nervous or defensive. On the other hand, if you hold open arms, you will make yourself comfortable and embrace others.

Distance

If you hold a person close, you can make yourself look pushy or put yourself in his face. At the same time, keep a distance away may mean that you don't care about what's being addressed or don't care about it at all.

You can study a lot regarding other people through body language. Many people show all sorts of thoughts in the way they push the body. Here are examples of how to discover other people's opinions:

Confidence

A comfortable person will always stand tall, holding eye contact sold while smiling at you simultaneously. The person can go even further with the hands when making gestures.

Tips for Reading Body Language

Eye movement, gestures, posture, and facial expressions are characteristics of human body language. American human behavior expert Eric Barker explains that it is best to look at "unconscious behaviors that are not easily controlled and may contain a message." Can anyone decipher this language? Barker reveals eight tips.

Use Common Sense

For the expert, analyzing the context is essential. Crossing your arms can mean adopting a defensive posture or even trying to deceive you. However, if it is cold or if the person concerned is sitting in an armless chair, the meaning can be different (much more straightforward and harmless).

Observe the Mime

Imitating a gesture or verbal expression may mean that the person is in tune with you. The act of agreeing with someone or something is difficult to fake, so the expert believes that the best thing is to think in these cases.

Nerve Energy

The other's level of activity can reveal your interest and enthusiasm for what you are saying. Research from the University of Manchester in England states that women shake their feet when they are interested in a man. Men, on the other hand, are tending to do so when they are nervous.

Consistency

Someone who reveals a fluid and consistent speech, emphasizing certain words, demonstrates control and concentration. Showing determination as they speak, these people are difficult to influence (and seduce).

Don't Care About Individual Signs

It is not possible to distinguish what a person means by body language through a single action. It is best to look at the actions as a whole, as two or three signals can help identify what goes in the other person's head.

Create A Reference

Uneasiness and the habit of always talking do not necessarily reveal any problem. However, something may be wrong when these people suddenly become calm.

Consider Your Previous Considerations

A judgment about another person will be affected if you have an initial impression, whether positive or negative. The inclination is to give the benefit of the doubt to someone you think is similar to you.

The Most Significant Thing Is to Focus on The Whole Context

Eric Barker argues that the ability to understand body language will increase when "understanding that body language is part of a larger context. Then you will begin to pay attention to other facets of interaction: voice, appearance, clothing, etc."

Chapter 3. Manipulation Through Body Language

After leaving the University heading to the golden paved streets of London, a man saw a woman who was also a model for the entire newspaper—at this point in his life. He hadn't met anyone attractive in the real world. Every day he walked and tried to be friendly, even cool, and afterward, he knew that he had just made himself a fool!

Why did trump react to her in that way? How was it that she was so successful? Have you ever addressed people with a lack of trust? And what about those other people in his class-his friend knew almost nothing about them as adults, he had never really been told, but he wasn't shocked by their performance.

They all had a certain aura about them that all these people could have a hypnotic effect on the people around them without opening their mouths. He wants to talk about this hypnotic body language today because it can help you achieve more and perfection without really having to do so much other than subtly alter your non-verbal communication.

3 Non-verbal keys for Hypnotic communication:

True Smile and Real Laughter

He can remember when his parents invited friends for supper when he was a child when his mother always told him to make sure that he was smiling and to show his teeth when the guests were arriving (I was never cheeky enough to grow, although the man was tempted). His mother knew that smiles produced positive reactions from people on an intuitive level.

This man speaks here of a real, genuine smile—a smile from the center of your body, which reflects joy. A natural smile makes the eyes and face wrinkle; insincere people smile with their mouths alone. Genuine smiles are often from the subconscious mind; individuals can sense, see, and feel real. A genuine smile implies you smile all over your face-your muscles move, your cheeks rise, your eyes shrink, and your eyebrows slightly go down.

Smile more, then. However, smile happily, fun, and joyfully. Smile in the future.

The explanation of why a photographer uses 'cheese' is because it's a term that helps you relax your face muscles. It often gives a crooked smile. How many pictures did you see that the smiles are cheese-powered and not authentic?

Professor Ruth Campbell, University College London, says that in the brain there is a "mirror neuron" which triggers the neurology responsible for the acknowledgment of face expressions and causes an immediate, unconscious mirroring response. The world smiles at you once you smile. In other words, know it or not, very often the facial expressions we see are unintentionally expressed.

So, if you smile more often than not-people around you smile more sincerely-it means, they feel better about you. You build for yourself and others around you a better immediate environment. How would you feel if you walk down the street and seeing someone with such an unhappy or cross face? Science has exposed that the more you smile, the more positive reactions people give you.

Would you smile more if you watch a funny movie with friends? Robert Provine found that Laughter in people in social situations was more than 30 times more likely than alone. He discovered that Laughter has less to do with jokes and funny storytelling and more to do with relationship building. Laughter creates a connection.

If you smile (a real smile) at another man, they almost always return the smile with a genuine smile that gives both of you and you genuinely positive feelings because of cause and effect. It creates a cycle of comfort: you smile, and you feel the perfect smile, and feel good, etc.

Studies show that most meetings run smoother, last longer, have better results, and improve relationships dramatically if you make a point of regularly smiling and laughing until it becomes a custom. He guesses you already knew all this-yet you smile a lot. Does recent research show that we smile 400% more-how as a kid often? Do you smile at the world today?

Confidence

The person was missing when he was younger, embarrassingly answering the receptionist.

I remember watching a documentary about a schoolgirl murdered in Great Britain. The girl's parents gave a press conference calling for help in the apprehension of the killer. It was the fall of the murderers. The way the father behaved at this press conference prompted the police to suspect him and to show him that he killed his daughter at last.

Many criminals are caught not because they have clues but because they are responsible, conscious of themselves, and lack trust. These feelings are sufficiently communicated to create suspicions.

When we are emotionally congruent and trustworthy, our body language is positive and expresses it to the world.

Psychologists advise us that by modifying our physical actions, we can alter our attitudes. Thus, adopting the physiology of trust can help you appear and become more trustworthy. When you are confident and hold your body that way more often, cause and effect mean having your body feel secure.

I recall reading a book a little while ago, and it taught you three great ways to build confidence with your body alone: first, he suggested that you be a 'front seater.' Wherever you go to movies, classrooms, meetings, and presentations, the back lines appear to fill up fastest, aren't they? Many people go back so that they aren't too visible. It often shows a lack of confidence in him. Start sitting up today, relaxed with other people's eyes, and build confidence.

Furthermore, making direct contact with the eyes tells you a lot about confidence. If someone avoids contact with the look, we might start wondering what's wrong with them or what they must hide. Lack of eye contact may indicate that you feel weak or that you are, in some way, afraid. Conquer this and let the person in the eye look—you don't have to stare hard! Just look in your eyes to tell them that you believe you are honest, open, confident, and comfortable.

Suppose you seem confident and think of yourself. In that case, the other person tends unconsciously to agree that there is something worth knowing about you-why should anybody else be if you aren't confident or feel good about yourself? It is implicitly conveyed beyond conscious minds, often with these sages' good feelings. David Schwartz gave the other great tip to walk 25% faster. This man knows that his father always told him to slow down when he was taken to football to see his beloved

Nottingham Forest as a boy because he was enthused and enthusiastic about their destination.

Psychologists link Slovenian stances and slowness to disagreeable attitudes towards oneself, work, and the people around us. But psychologists also tell us that by changing your posture and movement speed, you can change your attitudes. Body action is the result of mental action-and vice versa-as this man already said; cause and effect! The person with low morality is shuffling with little confidence and stumbles through life. Likewise, ordinary people are on average. You can see it, and you can hear it.

Confident people travel purposefully, they have to go somewhere important, and they will succeed when they get there. Open your chest, throw your shoulders back, lift your head, be proud of yourself, move a little faster, feel that your trust will grow. It doesn't have to be dramatic; just keep your body safe.

The Right-Hand Side of the Brain

Most people are right-handed, and as such, their thoughts and lives are processed on the right-hand side of their brain, and motor reactions and functional brain use reside on their brain's left-hand side.

Evolutionary psychologists debate it; most of them think we all have six raw emotions. All else is derived from these. Those six emotions are central: happiness. Surprise. Disgust. Fear. Rage. Anger. Sadness.

It's worth noting here that only two of them are good. If we are real, only one is guaranteed to be great to ourselves, isn't that? Following April's foolish day, he is reminded how much he enjoys' his surprise!

The vast majority of our thoughts in our minds are somehow negative. It is accurate, and bad things tend to stand out much more than our minds' good things.

So, if you respond to anybody's right brain, you may unconsciously associate yourself in the right mind with all those emotions. You don't want to do it.

If you first meet someone to use this knowledge in life instead, put yourself, so they have to look slightly right to look at you. See your right eye when you shake your hands. He believes that this picture is so much on his website's right-hand side. That in his rooms, the man places his chair so that his customers need to look correct when we communicate.

There are three powerful things to remember when improving your success and performance without opening your mouth.

Note that if you smile and smile with enthusiasm, if you behave with faith and connect with the right brain pieces, you start resonating far more gradually with the whole world.

Chapter 4. Uses of Body Language

Body language and self-esteem go hand in hand. It allows for a beautiful mechanism to observe and monitor how people behave and feel. Awareness of our body language is essential for becoming effective and persuasive communicators. Hence, there are several applications for using, reading, and changing body language.

Therapeutic Applications

Body language plays a significant role in counseling, NLP, and hypnotherapy. For psychologists, body language allows them to read their clients' emotional state and gives them a way to build rapport. Observing the client's body language can help the psychologist read how the client responds to a specific discussion or questioning line.

Body language speaks when we can't. Health care professionals have known this for some time. Many studies have been conducted in it and psychology academic studies for professionals, including modalities on body language.

Common issues which can be examined and treated through the use of body language include:

Bipolarity

Individuals with this condition suffer a chemical imbalance that leads to severe depression and the inability to make decisions.

They often have low self-esteem that accompanies this disorder, and it is incredibly challenging to understand effectively or treat correctly. The person with bipolarity can be taught to manage their daily situations. Considering the link between body language and emotion, they can also enjoy relief by being trained to use positive body language. It is a means for them to use their body language to persuade their emotions to stabilize and improve. For their families, body language reading is also an effective way to monitor their loved one's state and intervene before incidents happen. Depression can often go unnoticed, and people will rarely speak out about it. They are not likely to say: "I'm feeling depressed."

Low Self-Esteem

Many of us have suffered the disturbing effects of low self-esteem in one way or another. The first victim is our ability to progress in life. A positive belief in yourself is needed to convince the rest of the world to believe in you. People can be trained in positive body language such as open positions, eye contact, and lifting the head. It's a case of falsifying it until you feel it. With enough repetitive use of persuasive body language, you can even convince yourself that you are stronger than you believe.

Trauma

Survivors of trauma suffer from a loss of power, feelings of inadequacy, and loss of confidence. They also have the burden of guilt, where they hold themselves responsible for what happened to them. Whether the trauma is due to a violent act, these individuals' emotional state is reflected in their body language. Body language may have been positive and inviting before the incident. The person may display negative body languages, such as crossed arms, slumping, excessive facial touching, and nervous ticks such as repetitive movement. With effective counseling, their progress to recovery can be tracked through counseling and monitoring their body language.

Abuse

Abuse can be physical, emotional, and sexual, but whichever of these it is, there is bound to be an overwhelming sense of a loss of power. The victim may need to be convinced that they can regain their strength and that it is okay to trust people. Body language is extremely efficient in this regard. Helping these survivors of abuse establish strong body language will increase their sense of their strength. Suffering abuse is also linked to a loss of trust in people and the world around them.

By helping the abuse victim understand others' body language, they can be aided in evaluating the world and those around them regarding what they see, not what they fear. It is already great empowerment to the abuse victim, as they can become a participant in life again and feel like they can make informed decisions.

Self-Development

Being an effective communicator is one of life's excellent skills that will open doors and lead to the self's emboldening. Self-development programs often include body language modalities where the participants are trained in positive body language and assertiveness.

Group Dynamics

People can be classed into two groups: introverts and extroverts. Introverts, as we know, are those people who tend to thrive in one-on-one communications and prefer to spend more time alone, while extroverts are the life of the party and go through life with a the-more-the-merrier attitude. Introverts often suffer a form of depression based on social settings. They do not do well in groups. As a result, their communication within a group dynamic tends to fizzle. Yet, communication is a learned skill. As we learn the words, sentence structures, and grammar of a new language, we can also learn how body language works.

Depression

People suffering from depression tend to convince themselves that they are not worthy, that they are to blame for some usually imaginary flaw, and that everyone around them judges them.

People with depression sometimes think that everyone else has it right, while they alone are suffering. In creating awareness of body language, they can see the world in a more realistic sense and realize that people everywhere go through trying times and are not alone.

By learning to focus on using positive body language, they can also begin to manage their condition, encouraging well-being.

OCD

This condition is known for the repetitive behavior that someone engages in to make themselves feel in control of their lives. At the root of this tragic condition lies the fear of losing power and a profound distrust in themselves and others. In extreme cases, this can even encompass excessive washing of hands to remove imaginary germs and avoid people because people have germs.

People with OCD tend to have a very negative view of the world, and their only safety comes from their repetitive behaviors. Using body language, they can be trained to notice positive feelings in others and incorporate them into themselves. As they learn to project a positive self-image, they will feel their stress levels diminish, which will lead to a reduction of their anxiety-driven obsessions. When they feel more balanced, they will begin to develop trust in themselves and those around them.

Destructive Body Imagery (Bulimia and Obesity)

Low body image is a tragic and very destructive condition to suffer from. It goes with low self-esteem, lack of trust, feelings of abandonment, and severe depression. Bulimia leads the sufferer to obsessively lose weight, while obesity is a condition where the sufferer wants to fill themselves due to their emotional disabilities.

Both these conditions are associated with a loss of reality. These people begin to see the world not as it is, but as they believe it to be, and their world view is almost always negative. They eat or refuse to eat, to hide from the world and themselves.

Body language is a way to find a connection back to the real world. In reading the body language being projected by those around us, we can see that many individuals are just like us. We are not alone. Using positive body language is a therapeutic way to recover a sense of self that is realistic and beneficial.

The Biological Feedback Mechanism of Body Language

Due to our loss of trust in other humans, we often turn to animals for comfort and assurance. We read into what people do, what they say, how they say it, and how they react. A salesperson will do this on a second-by-second basis to monitor the client's body language and adjust their body language to match. Techniques such as mirroring, open position, advancing or retreating, and touching can be used to have an effect on the other person and monitor how persuasive we are on them. If they have begun to trust us enough, they will start to do something we want; in which case, we will trust them since they've done something for us. This endless, nonverbal loop is known as a biological feedback mechanism.

Training and Exercises

Some numerous academies and colleges strive to train people in body language detection and application. They mention facts and case-studies, what to do and what not to do; however, not many of them detail precisely how to improve your body language in a step-by-step way. When considering the activities and desired results, we suggest the following steps be followed:

Observe

Look at the world around you. Notice the people in it and how they interact with each other. Identify people in similar situations to those that challenge you. It could be someone applying for a promotion at work, asking a girl on a date, and even haggling for a discount. Each situation will use the same skills but in different ways. It all boils down to body language.

Practice

It will require some bravery, which is perhaps why people do crazy things in foreign lands where no one knows them. Find some friends, set up a hidden camera if you have to, or undergo obedience training with your dog. The goal is to place yourself in a situation where you can practice some of the skills and how they can be used.

If you feel overwhelmed, you can practice at home with a mirror. You might even find some online help with an online counselor who can perhaps observe you over Skype.

Evaluate

Look at the recording you made of yourself, or talk to friends who are helping you. Don't look at your awkwardness; instead, focus on each body language technique, how you applied it, and what the response to it was.

You may even give yourself a score or write down what you need to focus on. Recall rejoicing the successes, no matter how small. Then it's time to repeat step two, practice.

It may seem like an incredibly arduous task to learn body language, but it certainly is worth it. These skills of using space, posture, facial expressions, eye contact, gesture, and touch are vital to leading a fulfilling life that has less conflict and misunderstanding in it.

Chapter 5. Guide to an Effective Body Language

Research has shown that, when you are aware of your own body's happenings, you can manipulate it by training yourself to have control and even mold it to have effective communication. Further research recommends that you take some breathing exercises before going into a meeting or presentation. It will help you calm and have the ability to take note of your posture and gestures while on presentation. As you have noted by now, mirroring is a good technique. Always try to be keen on what the next person is doing non-verbally and copy that. It will help you turn out to be more effective in your communication with them. They will understand you better because this tunes your mind to communicate more truthfully at a place of relaxation.

However, you should be careful while shaping your body language. It is to ensure that the body language that you portray matches with what you are trying to present. A mismatch may bring confusion and may not be relevant at the moment. The person you are in conversation with my mistake you for meaning something else contrary to what you intended. The secret to having control of your body language is taking your time to learn it and being aware of your non-verbal cues as you apply what you know.

The Body Language That Will Help You Take Charge of Your Space

Effective management involves individuals being able to encourage and have a positive influence. In planning for a necessary appointment, maybe with your employees, management team, or partners, you focus on what to say, memorizing critical points, and rehearsing your presentation to make you feel believable and persuasive. It is something you should be conscious of, of course.

Here is what you should know if you want to control your position, at work, in presentation, or as a leader.

Seven Seconds is What You Have to Make an Impression

First impressions are essential in market relationships. When somebody psychologically marks you as trustworthy, or skeptical, strong, or submissive, you will be seen through such a filter in any other dealings that you do or say. Your partners will look for the finest in you if they like you. They will suspect all of your deeds if they distrust you. While you can't stop people from having quick decisions, as a defense mechanism, the human mind is programmed in this way. You can learn how to make these choices useful to you. In much less than seven seconds, the initial perceptions are developed and strongly influenced by body language. Studies have found that nonverbal signals have more than four times the effect on the first impression you create than you speak. It is what you should know regarding making positive and lasting first impressions. Bear in mind several suggestions here:

- Start by changing your attitude. People immediately pick up your mood. Have you noticed that you immediately get turned off after finding a customer service representative with a negative attitude? You feel like leaving or request to be served by a different person. That will happen to you, too, if you have a bad mood, which is highly noticeable. Think of the situation and make a deliberate decision about the mindset you want to represent before you meet a client, or join the meeting room for a company meeting, or step on the scene to make an analysis.

- Smile. Smiling is a good sign that leaders are under using. A smile is a message, a gesture of recognition, and acceptance. "I'm friendly and accessible," it says. Having a smile on your face will change the mood of your audience. If they had another perception of you, a smile can change that and make them relax.

- Make contact with your eyes Looking at somebody's eyes conveys vitality and expresses interest and transparency. An excellent way to help you make eye contact is to practice observing the eye color of everybody you encounter to enhance your eye contact. Overcome being shy and practice this excellent body language.

- Lean in gently the body language that has you leaning forward, expresses that you are actively participating and interested in the discussion. But be careful about the space of the other individual. It means staying about two feet away in most professional situations.

- Shaking hands. This will be the best way to develop a relationship. It's the most successful as well. Research indicates that maintaining the very same degree of partnership you can get with a simple handshake takes a minimum of three hours of intense communication. You should ensure that you have palm-to-palm touch and also that your hold is firm but not bone-crushing.

- Look at your position. Studies have found that uniqueness of posture, presenting yourself in a way that exposes your openness and takes up space, generates a sense of control that creates changes in behavior in a subject independent of its specific rank or function in an organization. In fact, in three studies, it was repeatedly found that body position was more important than the hierarchical structure in making a person think, act, and be viewed more strongly.

Building your credibility is dependent on how you align your non-verbal communication

Trust is developed by a perfect agreement between what is being said and the accompanying expressions. If your actions do not entirely adhere to your spoken statement, people may consciously or unconsciously interpret dishonesty, confusion, or internal turmoil.

By the use of an electroencephalograph (EEG) device to calculate "event-related potentials"–brain waves that shape peaks and valleys to examine gesture effects prove that one of these valleys happens when movements that dispute what is spoken are shown to subjects. It is the same dip in the brainwave that occurs when people listen to a language that does not make sense. In a somewhat reasonable way, they simply do not make sense if leaders say one thing and their behaviors point to something else. Each time your facial expressions do not suit your words. For instance, losing eye contact or looking all over the room when trying to express sincerity, swaying back on the heels while thinking about the company's bright future, or locking arms around the chest when announcing transparency. All this causes the verbal message to disappear.

What Your Hands Mean When You Use Them

Have you at any point seen that when individuals are energetic about what they're stating, their signals naturally turned out to be increasingly energized? Their hands and arms continuously move, accentuating focus, and passing on eagerness.

You might not have known about this association before. However, you intuitively felt it. Research shows that an audience will, in general, view individuals who utilize a more prominent assortment of hand motions in a progressively ideal light. Studies likewise find that individuals who convey through dynamic motioning will, in general, be assessed as warm, pleasant, and vivacious. In contrast, the individuals who stay still or whose motions appear to be mechanical or "wooden" are viewed as legitimate, cold, and systematic.

That is one motivation behind why signals are so essential to a pioneer's viability and why getting them directly in an introduction associates so effectively with a group of people. You may have seen senior administrators commit little avoidable errors. When pioneers don't utilize motions accurately on the off chance, they let their hands hang flaccidly to the side or fasten their hands before their bodies in the exemplary "fig leaf" position. It recommends they have no passionate interest in the issues or are not persuaded about the fact of the matter they're attempting to make.

To utilize signals adequately, pioneers should know how those developments will be seen in all probability. Here are four basic hand motions and the messages behind them:

- Concealed hands—Shrouded hands to make you look less reliable. It is one of the nonverbal signs that is profoundly imbued in our subliminal. Our precursors settled on endurance choices dependent on bits of visual data they grabbed from each other. In ancient times, when somebody drew nearer with hands out of view, it was a sign of potential peril. Albeit today the risk of shrouded hands is more representative than genuine, our instilled mental inconvenience remains.

- Blame game—I've frequently observed officials utilize this signal in gatherings, arrangements, or meetings for accentuation or to show strength. The issue is that forceful blame dispensing can recommend that the pioneer lose control of the circumstance, and the signal bears a resemblance to parental reprimanding or play area harassing.

- Eager gestures—There is an intriguing condition of the hand and arm development with vitality. If you need to extend more excitement and drive, you can do such by expanded motioning. Over-motioning (mainly when hands are raised over the shoulders) can cause you to seem whimsical, less trustworthy, and less incredible.

- Laidback gestures—Arms held at midsection tallness, and motions inside that level plane, help you—and the group of spectators—feel focused and formed. Arms at the midsection and bowed to a 45-degree point (joined by a position about shoulder-width wide) will likewise assist you with keeping grounded, empowered, and centered.

In this quick-paced, techno-charged time of email, writings, video chats, and video visits, one generally accepted fact remain: Face-to-confront is the most liked, gainful, and impressive correspondence medium. The more business pioneers convey electronically, all the more squeezing turns into the requirement for individual communication.

Here's the reason:

In face to face gatherings, our brain processes the nonstop course of nonverbal signs that we use as the reason for building trust and expert closeness. Eye to eye collaboration is data-rich. We translate what individuals state to us just halfway from the words they use. We get a large portion of the message (and most passionate subtlety behind the words) from vocal tone, pacing, outward appearances, and other nonverbal signs. What's more, we depend on prompt input on others' quick reactions to assist us with checking how well our thoughts are being acknowledged.

Strong is the nonverbal connection between people. When we are in a certified affinity with somebody, we subliminally coordinate our body positions, developments, and even breathing rhythms with theirs. Most intriguing, in up close and personal experiences, the mind's "reflect neurons" impersonate practices, yet sensations and sentiments too. When we are denied these relational prompts and are compelled to depend on the printed or verbally expressed word alone, the cerebrum battles and genuine correspondence endures.

Innovation can be a great facilitator of factual data, but meeting in an individual is the key to positive relationships between employees and clients. Whatever industry you work in, we're always in the business of individuals. However, tech-savvy you could be, face-to-face gatherings are by far the most successful way of capturing attendees ' interest, engaging them in a discussion, and fostering fruitful teamwork. It is said that if it doesn't matter that much, send an email. If it is crucial for the task but not significant, make a phone call. If it is extremely important for the project's success, it is advised to see someone.

Chapter 6. How to Persuade People

Persuasion is a deliberate effort to change or alter a person's opinions, beliefs, or attitudes toward an issue, situation, object, or person. It is usually achieved by transmission of a message which could be verbal or symbolic.

While persuasion could be used in a manipulative sense, it is, in an actual purpose, different from manipulation. It is because, when persuading a person, he/she is usually aware of your efforts at changing their point of view and willingly or reluctantly allows you to try. In this instance, the person listens and concentrates on what you are saying and then tries to rationalize your ideas with reality before then putting whatever conclusions they come to comparison with what they believed.

Your role in the entire dynamic is to state your reasons for the change you are prescribing, give illustrations and evidence supporting your views and try to convince the target of your advances that your line of action or advice is their best bet. The main goal of this is getting them to switch to a state of reasoning. In this, persuasion resembles manipulation because your goal is still to push the target towards an outcome that they might ordinarily not have considered right.

The success attained in persuasion usually depends on the target's preconceptions and their strength, their perception of the person sharing the new message or idea, their perception of the message or idea, and finally, their perception of the conclusion on offer. Upon outlining these reasons, it should be clear to you that the subject of your effort would probably possess ideas that are at least dissimilar. If not contradictory to yours and as such, the entire process would either hinge on your persuasion being very convincing or the target's ability to meet a compromise between the conflicting ideas that would majorly mirror the changes you want.

Below are six major theories that explain how the human mind absorbs and reacts to information. Knowledge of these would greatly increase the odds of persuasion if you could pinpoint it in your target.

The Attribution Theory

It concludes that people would either attribute actions and characters to people and objects, respectively, either relative to the context they are being considered in or according to their emotional disposition.

When they attribute using context as a guide, they are likely to come to decisions that consider the environment of origin and situational factors. Such is seen when a person refrains from calling a product inferior or calling a person insensitive. Instead of arguing that the product has been made from the best possible items available to the manufacturer. The person is merely reacting as he has learned from his childhood environment.

However, when considering their emotional disposition, they tend to believe that whatever is convenient for them is the only right decision or approach for every other person. Consider this situation:

You meet a person at an event or gather and try to start a conversation with them, but instead of giving you a polite audience, the person appears preoccupied with their thoughts or acts aloof. Angered or annoyed, you walk away, and when asked for an opinion on the person, you characterize them as proud, arrogant, or self-important.

In this case, the characterization you have concluded is based solely on your emotional disposition and does not consider the situation or possible problems the other person might have. The idea is not to determine whether you are wrong or right but rather to analyze how you are likely to process information about people and things. You might be right about the person.

Another situation is when you have been accused of doing something wrong, and you claim that your accusers have failed to see things from your perspective and are only interested in their point of view.

It is a perfect example of considering things as regards context. In this case, probably because the things said are negative, you'd notice the emphasis placed on contextual understanding of actions. There is also a minor hint of the dispositional thinking coinciding.

The Conditioning Theory

In this case, the person is likely to do things. It is if they are conditioned to look like their own decisions instead of coercion. It is mostly utilized in the advertising industry where commercials, advertisements, and billboards convey information that would provoke positive feelings in the population of interest. They then connect such sentiments to their products, making you feel that the work would bring such a feeling into your life since you are more likely to purchase their product, thinking that your decision was an independent one.

It is usually possible because we generally perceive things based on our emotions and are more likely to buy things because they make us feel good.

The Cognitive Dissonance Theory

Based on this theory, it is assumed that people tend to aim for consistency in their thoughts, attitudes, and decisions. It is the cause why most individuals create principles that they strive to follow. Most people also seem intent on reconciling the contradictions as much as they can until they feel comfortable. I would give two examples of this.

Example 1

You have an extreme and deep-rooted need for canned food, either due to the laziness of having to cook meals or the frustrations at having to wait in queues for food. Then you are told that such canned meals could lead to cancer, and you don't want to have cancer. But you also don't want to stop eating canned food. So, instead of stopping with the habit, you comfort yourself that millions of people like you have the same habit and never have cancer.

The cancer theory might be untrue, but your eagerness to dispute the fact or at least make the consequences seem less severe is your own way of changing your mind or making the facts you have just learned seem less important or true. It is one of the ways of dealing with cognitive dissonance theory.

Example 2

Imagine a criminal with a conscience. It is probably hard to envision, but they do exist. Their criminal tendencies are clashing with their tender hearts and causing a bit of discomfort in such a situation. Such a person is very likely dealing with his/ her problems by giving in to the rationale that a criminal and wealthy life far outweighs the benefits of having a clean conscience or right heart.

Again, I am refraining from judging whether such a rationale is sound but am more focused on the fact that the person seems to give in to a motivation that overlaps with most people's general aim to deal with his discomfort.

The Judgment Theory

This one is straightforward to grasp. It merely proposes that when faced with a new piece of information or idea, a person's reaction is dependent on the way he/she currently feels on the topic. What this means is that we're likely to accept something that resonates with our current belief, reject something that doesn't fit in with our beliefs, or stay indifferent to something never considered before.

Therefore, when attempting to persuade a person, it is better first to determine their views on the topic to gauge whether you'd be successful and if your effort would eventually be worth it.

The Inoculation Theory

The inoculation theory supports the view that even if uninterested before in two points of view, once argued for, you are likely to pick the dominant point of view and stick with it. Here is an example:

You have never viewed a soccer game in your life, but one day you are relaxing on the beach and happen to find yourself stuck between two diehard soccer fans who support rival teams. An argument begins about whose team is better and more dominant, and they both turn to you, presenting their points like you are a seasoned fan and, after some time, ask you to judge who's better. You obviously would pick the person with the better argument to not betray your lack of knowledge on the subject. If another individual were to pose a question to you in the future, inquiring about which of those two teams is better, you'd probably find yourself arguing in favor of the choice you made then, maybe even with some of the same points that were used then.

It is the power of inoculations; the most powerful initial idea always takes root first.

Narrative Persuasion Theory

From experience, I think we would all accept that stories have a more enhanced effect on perception and opinions than abstract advice. People's attitudes and views towards objects and others tend to change when they are told compelling stories of such subjects.

The theory simply attempts to explain the heightened effect that can have on people if appropriately utilized. In this, the listener feels transported, which significantly affects their perception of events, making them more pronounced and vivid than they might have been if they had been expressed ordinarily and abstractly.

The Psychological Perspective

Ordinarily, persuading people would be difficult without the ability to organize and present an argument properly. But if inexperience in any or both of this is coupled with an inability to understand moods and stances. Your task would be made many times more likely to fail.

The ability to instantly sense and recognize a person's stance on an issue is difficult, not to talk of performing the same trick on an audience. Because of this difficulty, most speakers who are attempting to introduce people to a new point of view always tend to ask questions that would enable them to gauge the audience's stance before moving on with their presentation.

After asking such questions immediately, they usually watch out for visible reactions from the audience members, maybe a smile to indicate a knowledge of the topic, sitting up to indicate interest, turning away, or sighing to indicate disagreement, boredom, or even a person willing to answer. These simple markers give you an idea of how your message may be received and help you map out a strategy of approach. It is also a useful tool as people express themselves more sincerely when they do not feel particularly in the spotlight. If you are unsure, do not refrain from asking a few surface questions to test the waters or, more aptly, to feel out the crowd.

It should also be noted that numerous people might give an adverse reaction to one-on-one persuasion and would start arguments to further their points. The moment you realize that your attempt to persuade a person has deteriorated into an argument, it is sensible for you to stride away. Very few disputes occurring outside law courts ever get settled. Engaging in one would be fruitless and time-consuming. That time is better spent elsewhere.

Chapter 7. How to Analyze People

Logic alone cannot help you if you want to understand an individual. To know how to analyze the non-verbal initiative cues given off by people, you have to give in to the other vital forms of information. In order to do this, you have to give up any emotional baggage or preconceptions like ego clashes or old resentments that might be stopping you from clearly seeing someone—the clue to this to receive information neutrally without contorting it and staying objective.

Whether you are trying to read your kids, your partner, co-worker, or boss, in order to do so accurately, you have to bring down some walls and surrender to any biases. You have to willingly let go of old ideas that can be very limiting. Those who can analyze other people properly are trained to read and analyze the invisible. They have learned to look further than where people generally look by utilizing their "super senses" and can access life-changing intuitive insights.

Analyzing People Effectively

In order to recognize how your mannerisms and actions can affect other people, you have to be able to comprehend the alterations between how you communicate with different people and how you act around them.

You need to note how all these people's different behavior affects you and how your actions make you appear to them. A suitable method of practicing this is by thinking about how other people might behave around you based on how they consider you in their lives. Maybe they act in a different way around you than they do around other people.

People You Do Know vs. People You Don't

How you see and behave towards someone is greatly affected by how well you know them or how well you need to know them. Your closeness to someone or distance from someone in the aspect of your relationships will define the things you need to contemplate when you are analyzing both you're and the other person's behavior while you are interacting with them. In the end, this will also help you determine how you want to make use of these insights in order to analyze what they are trying to communicate with you correctly.

Here are four examples to better elaborate on this concept:

1. You have an unstable relationship with your mother. Your relationship is long-term and intensely involved. You aim to find out the origin of the complication and fix your relationship with her. To do this, you first need to consider a few things: how she fulfills her needs, her points of concentration, comprehensive information about her personal life, the way she communicates with you, her impulses, preferences, and her body language.

2. You are in a relationship with your significant other for about a year. As it's starting to get more serious, you consider asking him/her to move in with you. The relationship is intimate and medium-term. Your first objective should be to think whether moving in together would be a smart move. You want to figure out how they might respond when you ask them the question. The essential factors that you need to consider are their past experiences and personal life, how they communicate with

you, their impulses, preferences, and body language. Besides, you also need to consider how they go about fulfilling their requirements, their points of concentration, and their drive. You can also acquire more insight by consulting family and friends.

3. You are thinking about sharing innovation for a business idea with a co-worker. You have a relatively superficial relationship with this co-worker, and it is medium-term. You want to observe their behavior to determine whether the two of you are compatible to work together and whether he or she would be a suitable business partner before expressing your idea. You want to figure out how you should approach them to get the best response. You need to observe the following factors: how they verbally communicate with you, their impulses, preferences, and body language. In addition to that, defining their concentration and having some insight into their past experiences and personal life. It could also be beneficial in this case.

4. In the initial process of meeting someone, you might ask yourself whether they are attracted to you. You are attracted to them; however, before expressing your feelings to them, you want to get to know them better. At this time, your interpersonal relationship with this other person is superficial and relatively new. In addition to that, before expressing your feelings, you want to be sure that you are correctly interpreting their signals if the feelings are not mutual. You need to pay consideration to a few things with your first encounters with this person. Some of the influences you need to contemplate are their preferences, how they speak with you, their body language, and how they convey themselves around you. You can subtly acquire some details like their history with relationships in your first few conversations and use that information to determine how you will act.

Techniques by Which You Can Analyze People

- Sense emotional energy.

Our emotions well express the energy or "vibe" we give off. Our intuition helps us register these. Some people help improve our vitality and mood, and it feels good to be around them. However, others can be draining, and you just want to move away from them. Even though this subtle energy is invisible, it can be felt feet or inches from the body. It is known as chi in Chinese medicine. It's a vitality that is important to health.

Methods to Analyze Emotional Energy

1. Notice their laugh and tone of voice—A lot about our state of emotions can be conveyed via our voice's volume and tone. The frequencies of sounds create vibrations. Try to notice how the tone of someone's voice affects you while you are trying to analyze them. Ask yourself whether their style feels whiny, snippy, abrasive, or soothing.

2. Notice the feel of their touch, hug, and handshake—much like an electric current, emotional energy is also shared through physical contact. Ask yourself whether a hug or a handshake feels confident, comfortable, warm, or off-putting. Is the other person's hand limp, indicating that they are timid and non-committal? Or are they clammy, meaning anxiety?

3. Notice people's eyes—People's eyes send powerful energy. Studies have revealed that similar to the brain that sends electromagnetic signals beyond the body, and the eyes do this. Take time and try to watch people's eyes. Are they angry? Mean? Tranquil? Sexy? Caring? Also, try to understand whether someone seems to be hiding or guarding something or are at home in their eyes, revealing their capacity for intimacy.

4. Since their presence, someone's company is like an emotional atmosphere surrounding us like the sun or a rain cloud. It's not essentially congruent with behavior or words but is the overall energy emitted by us. While you are trying to analyze people, try to notice: Are you feeling scared, making you want to back off? Or are you attracted by their social presence?

Listen to Your Intuition

Intuition is not what your head says. It's what your gut feels. It is the non-verbal information you can perceive beyond logic, words, and body language. What counts the most when you want to understand someone is who they are from within and not just their outer appearance. With the help of intuition, you can reveal a richer story by seeing further than the obvious.

Some intuitive cues you can look into:

1. Look out for intuitive empathy—You can experience an intense form of empathy when you can feel people's emotions and symptoms in your body. Therefore, when you analyze people, try to notice whether you are upset or depressed after an uneventful meeting or if your back hurts suddenly. Get some feedback to determine whether this is empathy or not.

2. Watch out for flashes of insight—You might get an "ah-ha" about people while you are conversing about them. It might come in a flash, so stay alert. If not, you might miss out on it. These critical insights might get lost as we tend to move onto the next thought very fast.

3. Feel the goosebumps—Goosebumps are amazing intuitive signals that tell us when we resonate with people who say something that we connect with or when they inspire or move us. It can also take place when you feel a sense of déjà-vu. Déjà-

vu is a feeling of recognition that you might have known someone before, although you haven't met.

4. Honor your gut feelings—During your first meetings, try to listen to your gut. Before you even have an opportunity to ponder about it, a visceral reaction already takes place. It conveys whether you are relaxed or not. Gut feelings take place very fast as a primal response. They act as your internal truth meter and convey to you whether you can trust someone or not.

Observe Body Language Cues

According to studies, words account for only seven percent of our method of communication. The remaining is represented by our voice (thirty percent) and body language (fifty-five percent). Stay fluid and relaxed while reading body language cues. Don't get overly analytical or intense. Simple sit back, be comfortable, and observe.

1. Interpret facial expression—Our feelings and emotions tend to get stamped on our faces. The deep frown lines convey Overthinking or worry. The smile lines of joy are depicted by the crow's feet. Pursed lips signal bitterness, contempt, or anger. Grinding teeth or clenched jaw are signs of tension.

2. Pay attention to posture—When you are trying to analyze someone's posture, ask yourself: Is their chest puffed out when they are walking, which is a sign of a big ego? Or do they cower while walking, which is a sign of low self-esteem? Or, is their head held high, confident?

3. Notice their appearance—When you are analyzing others, pay attention to their appearance. Are they wearing a t-shirt and jeans, indicating that they are dressed casually for comfort? A

power suit with properly shined shoes that are indicating ambition and being dressed for success? Or a pendant like a Buddha or a cross displaying their spiritual values? Even a well-fitted top with cleavage, representing a seductive choice?

Learning how to analyze other people accurately takes time and practice, and obviously, every rule has some exceptions. However, you can improve your abilities to analyze others, communicate properly with them, and understand their thinking by keeping these points in mind while building your powers of observation.

Chapter 8. Dark Psychology Secrets

There is a concept in the world of psychology, which is called the dark triad. The obscure triad is a set of three personality traits, namely, Machiavellianism, narcissism, and psychopathy. This group of three is labeled dark, owing to the usual malignant habits correlated with certain characteristics. The dark triad's dramatic opposite is the lighter triad, which is a topic and debate for another book in itself. Although the three traits depicted on the dark triad in their studies are distinct, it is seen that they also overlap. It indicates that with blurred boundaries, a person who increases the success on the dark triad exam will likely have all these traits present. It might be hard to tell, for example, where narcissism stops and where psychopathy begins.

Discussions about the Dark Triad concept were initially begun in 1998. Three psychology experts do it. They asserted that Machiavellianism, narcissism, and psychopathology occurred overlappingly in normal samples. Two psychologists by the titles of Williams and Paulus would later invent a name for this group, in 2002: the dark triad.

There have often been discussions and debates about nature's part in seeking to comprehend the dark triad's personality traits. To put it simply, psychologists, behavioral scientists, and researchers were keen to know whether born or bred are Dark Triad persons. Are we born stupid and manipulative, or have we become so as a consequence of the things that we grow up to be exposed to? According to various research done, it has been noted that a dark triad has an important genetic basis to it. That is, some born with such susceptibility to the dark traits of the triad. However, in terms of heritability, narcissism, or psychopathy, rank greater than Machiavellianism. That is when contrasted to a parent that ranks high on the Machiavellian scale, a psychopathic mother or father is more willing to switch the characteristic to their offspring.

The dark triadic characteristics have also been seen to be underrepresented in top-level management in reports that might not be really friendly to someone working. When the dark triad elements are unpackaged in the segments below, it becomes evident why this recognition may be so.

Dark Triad: Narcissism

A narrative is revealed in the Greek myths of a young man named Narcissus. Narcissus was indeed a hunter renowned for his striking, good looks. Narcissus did not have a time of day for them, given the adoration he got from his admirers and even forced others to take their own lives to show their devotion. While there are several varieties of Narcissus's story, all of them refer to him being extremely self-absorbed, which eventually ended up in him going to die mortality that was retribution for his selfish ways. Thanks to the story of that young man, Sigmund Freud first coined the term narcissism. Freud, aptly titled On Narcissism in his famous 1914 essay.

In the simplest terms, narcissism is the increased and compulsive self-admiration which a person has towards himself and his personal features. A narcissist is always easy to recognize since they quickly offer away their behavior and values.

Asking yourself if you have a narcissist in their life? Here is what you should look for:

- Narcissists tend to feel good and always have the ability to be entitled.

- Type-A perfectionists are also narcissists.

- Narcissists have an unflagging thirst for control.

- Narcissists don't have a sense of limits.

How Do Narcissists Control People?

Now that you realize how well a narcissist looks, you're possibly curious about what the narcissist is doing to manipulate you in your life. How difficult can it be to remember, after all, that someone is attempting to manipulate you? The response is, it can be quite challenging, particularly when this person conceals their acts as only searching for you. Many narcissists are typically very clever and can fit in their daily life without drawing attention to them. They could also be very creative and talented, and the allure that tries to draw you to them will usually be that. When you're out there going to look for a narcissist-shaped monster, you might not be going to look for that skilled and super artistic friend who's always having a solution to everything. And still, she might be the only narcissist in the life who just thinks about competing or who gets injured along the way.

Narcissists are also quite keen liars, in addition to using their mentioned characteristics to the best of ability. Narcissists conduct routine the skill of deception in its various forms in a bid to be the celebrity of every show. Deception is the way the narcissist throws you off reality, so they stay in control. In either scenario, they always exist in an altered world where they are good, and everyone else is inferior to them. Hence, deceit is just a means for them to draw you through this repetitive story where they are the principal character.

Dark Triad: Machiavellianism

Niccolò Machiavelli, sometimes referred to as the founder of modern social science, was a Renaissance-era Italian who favored loads of hats. Machiavelli has been amongst others a historian, politician, poet, humanist, author, and diplomat. Machiavelli composed his most popular work, The Prince, in 1513. In this book, Machiavelli defined and advocated the usage of unscrupulous methods for obtaining and retaining political influence. The word Machiavellianism arose from this work and its endorsements, that was used to refer to the kind of politicians and tactics Machiavelli mentioned in his book. This word was later coined by psychology researchers to define a psychological characteristic marked by a lack of sympathy and a drive to succeed at the detriment of others, be it by deception, coercion, or the flouting of traditional dignity laws and morality. A person who displays Machiavellianism is, in the simplest form, willing to do almost anything if it meant playing. Machiavelli is the purpose of why the ending phrase justifies the means that exist.

Most work has been conducted since the introduction of the word Machiavellianism in philosophy to ascertain what determines the people who score highest on a Machiavellianism test, better known as high Mach's. High Mach's have been found to tend to value power, money, and competition above all else. High Mach's put a very cheap cost on things like building a community, family, and even love. Among those that score low above the Machiavellianism index, better known as low Mach's, the opposite is accurate.

Dark Triad: Psychopathy

Psychopathy is a feature of temperament marked mainly by a loss of empathy toward others. Psychopaths barely experience empathy for others and will rarely feel remorse even when other people have been hurt. There are various psychopathy views, but many of them always seem to agree on the three primary features that differentiate a psychopath from any normal individual. These three traits include fearlessness, lack of restraint, and meanness that any other person would consider uncomfortable.

Psychopaths are brave and aggressive and are not reluctant to step into new terrain even though they could be in danger. Although most individuals are usually overwhelmed by these conditions, psychopaths should be coping with such scenarios as if doing their everyday activities. Psychopaths often have a high degree of self-confidence as well as social boldness that allows them to interact with individuals without the shyness or anxiousness that others may have. Often, whenever a gruesome crime has been committed, you could perhaps hear about the nature of the investigation and shudder while going to think to yourself: how can a man live with himself for doing so? It's business as usual for a psychopath to kill someone and then grab a sunny face up at their local restaurant. It is not to say that all psychos have killed somebody. Some psychopaths instead rendered their lack of sympathy and susceptibility to other transgression and crimes.

Psychopaths show impaired regulation of the instinct, so they cannot regulate their impulses. When a regular human gets a desire of any kind, they can sometimes bring it under control and speak out of that state themselves. For instance, if you're having to deal with an irritating colleague who just won't be shutting up regarding their forthcoming bridal shower, you'll probably be able to combat the desire to slap them in their face. On the other hand, a psychopath will often be resolve by instinct and will react without giving it a second thought about the cost of everyone's decision. Psychopaths are susceptible to snapping in a simple way. Even one gets injured as they pop.

Common decency, when dealing with others, demands a certain level of decorum and kindness. It is not something of concern to psychopaths. While the majority of the people are worried about kindness and caring, the nicest person in the room will have no issue being a psychopath. Based on the situation in hand, they might be dramatic or execute about it.

The Dark Triad Practice

The dark triad test gauge how one score, as for the three practices of narcissism, Machiavellianism, or psychopathy, is concerned. The test is sometimes used in various settings, and by law courts and police in particular. The dark triad test is also used by corporations to gauge their employees. The primary reason the dark triad method is implemented is to assess an entity's personality characteristics and likely forecast their actions to prevent unsavory behaviors. It was noted that people who score high on the dark triad study are more likely to cause problems and social distress, whether in the work environment or even in their employment state. At the same time, these people will also likely have an easy time to attain leadership positions and gain sexual partners.

The dark triad test asks you to address a series of questions on a range of subjects like how you think about others and yourself, how you maintain track of details you could use to harm someone, and your general opinions on existence, death, and social experiences, among others. The dark triad test may be a nice way to gauge how you perform on the dark triad scale when self-administered. The dark triad test might not be very precise when administered by law courts and police as the respondent may purposely alter their answers to make them look better than it actually is. It is a primary drawback of the triad check in the night. If you're willing to take the dark triad exam, there are some online places where you'll be able to complete a study in minutes. Be careful to take the test results too personally—sometimes, the justifications you give are based on the kind of day you are taking and not on the type of person you are being. In any event, if you recognize yourself as a respectable human being who always respectfully treats others and never harms others, then you shouldn't worry very much about what an experiment says about you.

Chapter 9. How to Defend Ourselves of Dark Psychology

First, Identifying Them

By now, we have examined the foundations of dark psychology, the psychological profiles that make up the Dark Triad, typical forms of manipulation in relationships, and how manipulation has manifested itself in society's institutions.

First, remember that simply because you are not currently in a personal or professional relationship that could be defined as manipulative does not mean that you are free of all danger and concern. Predators have had to learn the hard way to live and achieve success using cold and calculating psychology from which they truly do not ever get any rest.

Imagine being injured in a serious accident and losing the use of one or more of your limbs. Regardless of how much you would prefer to have the use of that limb back. You will be forced to find some way to adapt. Emotional predators do the same thing. But because their injuries are invisible, and because of the business world's competitive nature, they sometimes hold an advantage over us if we fail to maintain vigilance.

Emotional predators can blend into the normal landscape because it is easy for them to go through daily living motions. They truly do not care if things don't work out because they have no value for their relationships or the things that society has established as having value.

Consider that the serial killer Ted Bundy worked on a crisis hotline while he stalked and murdered young women. He appeared successful, outgoing, handsome, and well-adjusted, but he was not. Or consider that the serial killer John Wayne Gacy, who murdered and buried in the crawl space beneath his home, almost 40 young men and boys, spent his days running a construction business, held fundraisers for local political leaders, and entertained sick children.

It may seem nauseating, especially with these extreme and dramatic examples. Still, for the emotional predator, society's important responsibilities are less a source of personal and professional satisfaction and fulfillment and more a perfect cover for their predatory addiction. As a result, you may find it helpful to develop some habits that will help you learn to classify some of the significant signs of emotionally predatory behavior.

Not everyone's life is perfectly organized or compartmentalized. Often environments and the people in them cross boundaries. Often in our daily lives, we wonder where things may have gone wrong. Quite often, the answer may be that we are trapped in a relationship with an emotional predator.

Regardless of the environment in which you meet people, you should always maintain a vigilant lookout for any of the following telltale signs of a predatory personality:

- Pathologically selfish people. They may go through friendship and love motions, but their emptiness is apparent when they fail to initiate social outings or when all encounters leave you feeling exhausted and drained.

- Emotional predators may offer lots of charm and flattery, but if there is a lack of substance to your interactions with them, you can be sure the compliments are probably false, too.

- Predators will exaggerate their accomplishments and even lie. If you call them on it, they will refuse to take responsibility or admit that they are wrong.

- A date or outing with an emotional predator may always be a high-stakes adventure. If you never seem able to engage with them simply over a cup of coffee and have a happy and fulfilling encounter, you may be dealing with a predator.

- Predators are bullies by nature and use anger as their primary means of communication. Avoid people who demonstrate a tendency to humiliate people or challenge anyone who seems to have more power or success than they do. Predators also use insults and putdowns to build themselves up. You may notice this kind of conduct directed at other people when you are out with a predator. For example, if you are at a café or restaurant, a predator may try to impress you by insulting or humiliating the staff.

- Predators are manipulative, which they often show by making promises and then not keeping them.

- Because predators lack a conscience and do not understand that their abusive behavior should make them feel bad, a telltale sign, maybe anyone who boasts about committing abusive actions or crimes

- Predators may also display parasitic behavior. If you are tangled with someone who is excessively lazy and uses you, you should find a way to end the relationship.

Guidelines

Of course, identifying the signs of predatory behavior is only half the battle. The other half is discovering a way to resolve the conflicts and repair the damage that inevitably follows in the wake of an encounter with an emotional predator.

The following are some general guidelines. Some of the tips are meant as suggestions that you should implement on a daily basis. They should become new habits that will now be part of your daily routine. It is important not to regard these tips as chores or burdensome or a diversion or interruption of your normal life. Think of these suggestions as your own personal investment in your daily professional development.

Suppose a virtuoso musician who plays violin for a symphony wants to stay at the top of his profession. In that case, no matter what else he does, one thing must remain constant: daily practice and a constant effort to stretch his repertoire by seeking out more challenging pieces, finding new forms of expression, and adding new skills to his resume. Consider a university professor in any department—being hired into a tenured position is the only beginning. The "publish or perish" mentality will soon take hold. He will find that continually refreshing his professional assessment of his area of expertise is as much a part of his daily professional routine as the more mundane tasks involved in classroom lectures.

So, it is with life in the modern world. To maintain a position of success and happiness and fulfillment, we must think like any gifted performer or professional. Constant vigilance and the continual addition of new weapons to your arsenal to fight the war against the growing threat of epidemic levels of emotional predation will keep your calendar full.

Buy a notebook, start a new spreadsheet, create a new folder in your favorite browser's bookmarks tab, and clear off a shelf on the bookcase in your office. This effort in your life can be just as much a passion and an investment in your success and happiness as the money you spent earning your college degree or the time and effort you spent building your professional network.

Most importantly, as we move down the list of tips for dealing with predators, remember that it is not unusual to find that recovery from such encounters, in some cases, may take years. Though the first step of dealing with a predator is ensuring, they are no longer physically present in your life. This step is not always easy to accomplish. And once you achieve this goal, actually repairing the damage they have caused may keep you very busy for some time to come. But relax—though the damage inflicted by emotional predators can grow increasingly worse over time, so the benefits of successfully dealing with these incursions can have increasingly beneficial returns over time.

Here are some suggestions:

Conduct A Self-Inventory

From time to time, read the details in the types of character traits that make people more susceptible to emotional predation. Look within and be truthful with yourself about your weaknesses. Don't do this as an exercise in self-abuse, though.

Consider that an emotional predator approaches you with only one goal in mind—to destroy you. You may not be entirely willing to examine yourself in an unflattering light, but an emotional predator who has made you a target may not have time for anything else.

Be Cautious

Whenever you are meeting new people, whether romantically or professionally, guard your personal information

Resist Projection and Gaslighting

When you encounter these environments, remind yourself that the goal is to defeat all genuine efforts to establish accountability.

Keep A Journal

You don't have to be eerie about it, but respect yourself enough to seriously take your personal and professional aspirations. Write down your thoughts and concerns at the end of the day, even if you can only manage a few sentences. The blank page will never pose the kind of threat to you that an emotional predator may.

By getting your complicated thoughts out of your head and on paper, you have unburdened yourself in a means that is most useful to you. A predator knows you have this need, and their willingness to listen may be designed as a trap.

Go "No Contact"

If you are in a professional or private relationship, and notice any of the signs of emotional predation, take steps immediately to end the relationship. Sometimes that may mean not replying to text messages, voice mail messages, or email messages. The predator may not like it and may react angrily, but if you try to enter into a negotiation or debate, you will be playing into their hands. Just say that you have decided not to respond any further, then stick to your plan.

Going "no contact," and in the modern world with all its digital communication, is a valid and acceptable tactic. If the predator continues to harass you, keep notes, and document their abuse. You may need to use it later if law enforcement becomes involved. Screenshots, text messages, email messages, and voice mail messages should all be saved and kept in a folder.

Get Help

Recognizing that you are in a relationship with a predator is the first step to escaping the relationship. Rescuing yourself must become your first priority. Remember that you will require professional help to solve this problem. If you are unsure how to proceed, take ten minutes out of your day, find a quiet place, and make a phone call. Don't worry about being perfect or feeling awkward. Professionals expect you to be at a loss and will know how to help.

Find A Support Network

You may need to seek the support of the law enforcement authorities. If you believe things are that bad, you are probably right. Don't let yourself be bullied or intimidated. As with a psychologist or helpline call, making the first call is the most important step. Even if things don't go exactly the way you think they should, by informing the local authorities, you will have placed yourself in a better position

Reinvent Yourself

Remember that as a victim of emotional predation, you will no longer be the person you once were and will have to restructure your thoughts and approaches to life.

Cheer Up

You have taken the first step toward defeating the predatory influences that have brought the dark cloud over your life. It is the first day of the rest of your life, not the last day of the life you used to live.

As you move forward with your new awareness of your surroundings' nature, the world may become a less intimidating place, and you will once again find the joy and happiness that seems to have been lost for so long.

Chapter 10. The Power of Hands

The hands have a power that we do not know at all, and we ignore it. They send a huge number of messages that most people can't get.

Hands have always been used in conversation, and their meaning has changed countless times over the years. An example is a handshake.

The act of shaking hands finds its roots in the past. When the ancient tribes met, they used to show their palms to show that they were not hiding anything.

The Roman Empire instead used to tighten the forearm, so both people were sure that the other did not hide anything under his sleeve. It was done because, in those days, it was normal going around with a knife under the sleeve and being safe. They adopted this habit.

But like all the customs that have passed from generation to generation, the one used by the Romans has turned into our handshake.

This gesture for us is used in a myriad of different situations. They were ranging from the classic greeting with friends to a handshake to establish a working agreement between two large multinationals.

Even in Japan, where the classic greeting has always been the bow, the handshake is widely used today.

The fact that it is now a widespread gesture does not mean that it is simple to do. Behind the handshake, there is a real-world of domination and submission.

Still, in ancient Rome, two people greeted each other with an arm-wrestling handshake, I define it.

In other words, it was not common to shake hands as we do today, but one person took the hand of the other from bottom to top and created the shape of a sandwich, so to speak. The most powerful person dominated the other.

Nowadays, this practice is not used, but the person you win always exists while handshaking. There are three different types of endings for a handshake, which are:

- Dominance

- Submission

- Equality

These attitudes are perceived at the unconscious level, and our body processes them in a particular way, and each of these can decide in which direction the conversation will go.

An example I can give you is that of a study done on some company managers.

Male or female makes no difference.

It has shown that 89% of them use the dominant handshake and always hold out their hand first so that they can control the handshake accurately.

The exact opposite is a submissive handshake. In this case, the person puts his hand palm up, granting the other person dominance. A bit like dogs do when they lie down and put their bellies to the sky.

You can use this handshake if you want your interlocutor to feel in control of the situation. You can use this squeeze when you go to make excuses, for example.

On the other hand, when the two people are in a position in which both want to turn the other's hand to dominate, a "bite" is created. It causes the people to be equal, and neither of them, in the end, gets the better.

So, if you want to create an equal relationship with the person in front of you, avoid him turning your hand, but most importantly, use the same amount of force that he uses.

Now let's use hypothetical numbers. If he applies a force of 9 out of 10 to the handshake and you apply one of 7 out of 10, you will have to increase the strength, or you will be dominated. The same thing you will have to do in reverse if you don't want to dominate.

In short, if he applies a force of 5 and you of 7, if you do not want to be seen dominant, you will have to lower the power of your grip forcefully.

But now I'll tell you a trick to never let yourself be dominated. Not even if you were to meet the president of the united states.

Indeed, with this technique, you will always and I repeat, always dominate the other. Always if in that situation you want to do it.

The technique is called "disarming the doers."

The technique consists of putting the arm outstretched with the palm facing down to not leave any escape for your interlocutor, and he will have to turn his hand and put himself in submission forcefully.

From that moment on, you can do whatever you want. You decide whether to dominate or be equal, but it will be very difficult for him to bring the situation in his favor.

A bit like it happens in games when you are three points above your antagonist, and the game is about to finish, he has to do a miracle to win; indeed, the options for him are to draw or to lose.

If you occur to find yourself in the situation where a person holds out his hand as described above, there is something you can do to reverse the situation.

Step forward with your left foot and make sure to bring his hand vertically. This practice is not simple because we tend to advance with the right, but you will see that it will come more than natural to you with a little training.

If you really can't take this step, there is another way to save yourself from domination, and that is the double catch.

When the other brings you to palm up, you use the other hand, free to return the hold to a tie. So right now, you are using two hands while he is using just one.

Staying on The Left Is an Unfair Advantage

During a handshake, your position is crucial, and staying on the left helps a lot if you want to dominate.

It happens because, on the right, you have no control over the situation, while on the left, you can actually do it.

Kennedy liked this technique very much, even if at that time, nothing was known about body language; he already applied it by intuition.

If you go to see all the photos where he meets with leaders and famous people, you will always find him on the left with the double grip.

A striking example of how Kennedy was a phenomenon with body language is when he won the Nixon election.

At that time, it was renowned that the people who only heard the two politicians' speeches were convinced that Nixon had won while those who watched the scene agreed otherwise.

It led Kennedy to win the election. Pretty important this body language, isn't it?

However, going back to the speech above, if you are on the right of the photo to be able to have an equal situation, reach out to force him to shake your hand as you want.

To conclude, I give you a summary.

Few people know what an impression they can make on a stranger, even if they are conscious of how vital it is to yield a great starting point in a conversation.

Take some time to experiment with the various handshakes with perhaps friends, relatives, or work colleagues to get familiar with it. During important moments you will know how to behave correctly.

Chapter 11. Body Language of a Child

First of all, we must dialog about why it is different from reading a child's body language. The first reason is that they are young and have not yet learned to control their emotions. If a child is sad, they cry. If a child is happy, they smile. If they are angry, they yell and make mad faces, and if they are embarrassed, their cheeks turn red, and they hide their face. Some children might even decide to tell you about the emotions that they are feeling. Children are new to the world and have no reason to hide the things they are going through.

Because of this, children have body language that is extremely easy to read. They do not distinguish how to control their emotions, so they always show how they feel. If you read the emotions of a child, you are reading what they truly feel.

Another important thing to note about reading the body language of children is that since they do not yet know how to hide their emotions, they also are unaware of the body language signs that they portray. They are not capable of sending the opposite signal of how they feel like adults are.

Their lack of awareness of their own body language can also make it easy to spot when a child is lying. A child might try to hide the truth through their words, but they do not have the wherewithal to think to conceal it in their body language as well, often allowing tells to slip through that they are not telling the truth or are omitting part of the truth.

For example, my sister has a five-year-old daughter who likes to sneak chocolate chip cookies before dinner. My sister always checks the Chips Ahoy package right before she starts making dinner, so she knows when a cookie is missing. She will still ask her daughter in the hopes that her daughter will confess on her own. My niece will most of the time try to lie about it (my personal favorite being when she claimed her father ate the cookie). However, no matter how convincing she might think she sounds, she has one big tell that she is lying: a huge smile plastered on her face. Because she thinks she is getting away with something so mischievous, this smile appears on her face as she is so proud of her deceit. When it is clear that she will not get away with it, this smile is usually replaced by another tell, i.e., her hanging her head while looking at the floor because she is ashamed at having been caught.

Everyone has the physical tell that gives them away when they are lying. Fortunately for parents, guardians, and teachers, children are unable to hide their tells until they are older and have more experience both with lying and reading their own body language.

Today that we know how simple it is to read the body language of a child, let's look into how important it is to pay attention to the signals a child is conveying. Whether you are around children a lot or not, you need to be able to read a child's body language so that you can do your part in ensuring that our children are healthy and safe. Like reading an adult's body language can help us determine if they are in a dangerous situation, we can also use a child's body language to determine if they are in any danger. The unlucky truth is that we live in a world in which people will abuse, kidnap, and otherwise harm children. We want to help children out of such situations, but it can often be hard to tell when something suspicious is going on.

Reading a child's body language can help us determine if there is more to the state than meets the eye. Because young children do not know how to control their body language, any discomfort they feel around a specific adult will manifest in such ways as to how they hold themselves around this person. For instance, if a child exhibits such body language as standing stiffly, hunching their shoulders forward to make themselves smaller, or avoiding eye contact with everyone, including the adult they are with, it could mean that they are afraid of something. If they flinch whenever the adult that they are with reaches over to touch them, it could very well mean that this fear stems from someone hurting them on a regular basis, most likely this adult. Also, suppose they refuse to initiate physical contact with this adult while still never wandering any significant distance from them. In that case, it could mean that they are afraid to have any intimacy with this adult and of doing anything to anger them.

Mind you, none of this is a reason to call the police or Child Protective Services on someone. After all, there are multiple interpretations of any given body language. Standing stiffly, hunching their shoulders forward, or avoiding eye contact, for example, could just mean that the child is not comfortable in that particular environment or with strangers. Flinching and avoiding initiating physical contact with the adult could indicate, rather than fear, that the child has a problem with physical touch overall or that they are mad with that adult for some reason. Not wandering far from the adult, even though it is natural for a child to want to explore, could simply show that the child is well behaved or not particularly comfortable with checking out their surroundings on their own.

Like with all body language reading, what a child's body language means often depends on the context. If you know the child and adult personally, it can be easier to determine what the child's body language means. If they are complete strangers, it will be trickier. Nevertheless, spotting such body language in a child will help you to be on alert so that if suspicion arises that the child is being abused or has been kidnapped, you will be ready to take action.

Reading a child's body language will also help you to be there for them emotionally. If you have a child or take care of a child for large amounts of time, they will consider you their support system. They need you to help them learn about their lives and the world around them. It includes learning how to handle their emotions.

Sometimes, a child might have an emotion that they do not yet know how to explain. They may express this feeling through body language but still feel frustrated when they are unable to put their experience into words.

As an adult who distinguishes how to read body language, you can help in this situation. You can read the nonverbal cues that the child is portraying and use them to help the child express his or her feelings verbally. It will help the child learn about their feelings and more about who they are. It will also help the child grow up knowing that feelings are healthy and that it is okay to share your struggles with those close to you. If you can help your child in this way and teach these things to your child at a young age, they will have significantly fewer emotional struggles over the course of their life. This understanding is important to any adult who deals with children, such as doctors, teachers, and parents dealing with other kids, such as their children's friends.

Parents and caregivers need to teach their children how to express their own body language. Still, they need to teach the kids about simple body language reading techniques. You might not want to call it body language reading to them because they either will not understand or will think the topic is boring, but this skill must be taught to children in whatever creative way necessary.

You might wonder why I believe it is important for children to be able to read body language since it is a science-based topic that can be complicated at times. We will explain why this is important now.

First, if your children understand that nonverbal communication has just as much meaning as the words they speak, they will understand the people around them at a new level. Take their time on the playground, for example. If they ask a friend to play with them and the friend says no but is looking at the ground and has another child staring at them as if to tell them not to play with the child, they will know that there is more meaning behind this situation. They will either be able to speak up for their friend and encourage them to do what they want. Or they will be deprived of worrying about what other people think or be able to walk away without feeling offended because they know that there was more to the conversation than a simple denied request to play. It might even be a sign that the friend was bullied away from playing, and your child will be able to express to a trusted adult what they saw.

Also, think about if your child sees a classmate that is not saying much when they usually talk all day, every day. If your child is aware of the body language of the people around them, they might notice this difference in behavior and ask the child what is wrong. It could make a profound change in the said child's day.

You might even consider the friendships that the child already has. As an adult, you know that being able to read simple body language allows you to have better friendships. It makes sense, then, that the same is true with friendships among children.

Your child will also be able to avoid being a bully better if they are aware of their own body language. They will understand that actions like rolling the eyes or walking away from someone when they are talking to them hurt just as much as mean words. They will understand these actions and avoid them to be nice to the people around them when other children might accidentally hurt their friends with actions like these without knowing the consequences.

When a child knows body language, they are able to make sure that their friends are comfortable with them. If the child sits close to a friend, they will be able to tell if the friend is okay with close contact or not. If the friend is not tolerable and shows signs of being uncomfortable, the child will know that the right thing is to move away.

A lot of these types of body language are things that children learn through real-life experience. The only problem with this is that real feelings are getting hurt if they are learning in real life, and real friends are feeling uncomfortable. The sooner a child acquires these skills, the sooner they can use body language to their advantage.

Chapter 12. Reading People Through Their Mind

How often have you heard somebody instruct you to simply say something because they can't guess what you might be thinking?

Things being what they are, this is just half evident. The individual disclosing to you this may not know, yet they are positively equipped to guess what you might be thinking. They do what desires to be done in a more unpretentious way than they see.

A great many people can actually figure out how to guess thoughts with preparing, time, center, and a specific arrangement of abilities. It isn't something just mystics can do.

Even though clairvoyants do have the best possible ability, preparing, and "gift of perusing," it is surely something that can be figured out how to a degree.

Before I give you how everything people can figure out how to understand minds, it's essential to know some foundation data on mind perusing.

When you understand the science and the brain research behind psyche perusing, you will see that it is a reachable undertaking for anybody with the assurance to learn. And there are also a few deceives you can use to give the hallucination that you understand personalities.

Those stunts become considerably more helpful when you know the reality behind brain perusing.

Characteristic Mind Readers

The motivation behind why anybody can figure out how to guess thoughts is because we do it as of now.

Even though our suspicions are often off-base, it's not because the cycle of psyche perusing comes up short. We can reflect on the considerations and sentiments of people we cooperate with.

In any case, we often center our response around what we figure they will do rather than what they are revealing to us they will do. We often observe somebody's outward appearances and body language and effectively surmise that they are discouraged, debilitated, cheerful, irate, or content.

However, what happens when somebody has a decent poker face? Would we be able to at present guess thoughts without these visual signs to guide us?

Of course.

Required Skills

Truly is don't take those numerous aptitudes to understand minds. All you want is the drive to learn and the eagerness to incline toward your instinct when it mentions to you what somebody is likely reasoning or feeling right now.

You'll clearly require some training before your capacities easily fall into place for you. Be that as it might, you do not take to purchase a precious stone ball, an exceptional deck of cards, or an abnormal outfit to guess the thoughts of others.

You should have the option to free your brain from all interruptions before you endeavor to guess what someone might be thinking. For certain people, this will be the ability that sets aside the most effort to create.

Maybe you could take some yoga classes. Not exclusively will they assist you with centering your psyche and your energy. However, they will likewise give you some quality adaptability and exercise.

If you are searching for additional assets, these may help.

Tips for Beginners

If you need to figure out how to understand minds, you can follow some basic hints to kick you off. Widely acclaimed mystic Kiran Behara created these tips.

Behara's customers incorporate the absolute most extravagant and most well-known appearances in amusement and Broadway. You should begin by rehearsing these tips for your loved ones.

You should see snappy outcomes; however, it will take some time and practice to guess total outsiders' thoughts.

So here we go.

Open up Your Spirit

Notwithstanding freeing your brain from all contemplations and stresses, you should open up your energy to the people and potential outcomes around you. Try not to consider anything.

You simply need to be available at the time. Your psyche and soul should absorb the energy radiated by the people and things around you. Yoga is incredible at showing us how to do this.

However, you can learn it all alone at home in the quietness of your room.

Simply ensure people will disregard you while you start to center your musings and energy.

Seeing and Not Seeing

Free your thoughts.

Take a couple of seconds to perceive the individual sitting close to you genuinely. Make a psychological depiction of their facial structure, hair, eyes, stance, body language, and other subtleties.

However, you likewise should see everything else around that individual.

You must have a psychological segment that isolates the individual's characteristics and the other things that don't have a place with that individual. Separate the individual from the seat they are sitting in or the divider behind them. These things must be envisioned with a particular goal in mind so you can feel all the energy being created around you.

Zero in on the Person

Presently you need to restore your concentration to that individual's face. Look at them legitimately without flinching for around 15 seconds. Try not to gaze too long, or you may intrude on the energy by causing the individual to feel awkward. Following 15 seconds pass, you will need to turn away.

Make a psychological image of their face and their eyes. What does their energy feel like? Sit peacefully now as you let the considerations and sentiments of that individual fill your psyche and your spirit. You have now really begun the cycle of psyche perusing.

Start a Conversation

It is the place you will reveal the contemplations and sentiments of the individual. You can pick any topic you like for discussion. Get some information about their work or their home life. The considerations that come racing into your own brain might be the same musings crossing the other individual's thoughts. You could promptly mention to the individual what you accept they are thinking. If you have a clad memory, you can store these contemplations for later to summarize your whole impression of their considerations in these meetings.

The key is to invite any contemplations that enter your psyche now. Regardless of whether those considerations are dull and irksome, you need to give the individual a precise perusing of their contemplations. So as to do that, you should keep your brain open to each chance.

Possibly you didn't have any sign before that your brother is discouraged, and this disclosure harms you. However, your brother should realize that you're presently mindful of his battles. And now you may have the option to support him. The capacity to guess thoughts will give you a ton of intensity; however, it's a brilliant force if you use it shrewdly.

Other Tips

There is other advice you can use to increase these tips. When you increment your capacities to zero in on others' musings and sentiments, you can use more tips to give you an away from what goes on in other people's psyches.

These tips will build your odds of progress and knock the socks off of your companions, family, and outsiders you meet in the city.

Passionate Intelligence

If you realize the individual you're conversing with, you can inquire whether they feel similar feelings you're feeling. You'll show restraint toward this. Numerous people aren't truly adept at naming their feelings. They may feel furious when they're truly simply worried.

They could feel apprehensive when they're simply prepared to proceed onward to something different. If the individual you converse with concurs with the feelings, you sense, inquire whether they can sort out any reasons they may be feeling thusly.

Finally, you can start to offer recommendations on what they ought to do close to intensify or diminish these sentiments. They will be flabbergasted at your premonition and acknowledgment.

It may sound more like psychiatry than mystic brain perusing. However, it's one of the key approaches to build up your regular aptitudes.

Create Keen Listening Skills

What do all incredible communicators share practically speaking? They should be acceptable to audience members.

When somebody talks, be totally at the time with them. Try not to tune in for having the option to react. Tune in to the other individual with the goal that you can measure and understand all that they are stating. Yet, you should likewise tune in to what they're not saying too. If somebody isn't anticipating the remainder of their day, there must be a purpose behind that. Cautious listening will assist you in revealing those reasons and make them known to the individual. So as to succeed, you'll have to figure out how to listen more than you talk some of the time. Listening is the manner in which you find out about people and their feelings.

Try Not to Ignore Emotions

The explanation people need compassion today is because they decide to. Throughout each day, we are told to disregard our emotions to complete our work. Then put on a solid face for the world.

The more we overlook our emotions, the snappier they disappear. Rather than considering the new email from the chief or what you'll have for supper later, consider how you feel. As per proficient mystics, the more you can react to your own emotions, the more you will have the option to peruse and react to other people's sentiments and considerations in your life.

Guessing thoughts is something everybody can do and isn't only something for proficient mentalists and mystic peruses. You probably won't experience a lot of accomplishment from the outset, yet you can accomplish incredible advancement with training. Absolutely never utilize your new capacities to increase a bit of leeway over another person.

If you can peruse their feelings incredibly well, you may have the option to utilize that to get your direction. Utilize your capacities to help people. Brain peruses can be extraordinary companions and emotionally supportive networks for people who simply need to vent.

You have probably thought about how things would be if you could guess other people's thoughts. A few individuals utilize their instinct for this, yet if you are not all that discerning, there is just a single decision left: reckoning out how to peruse people's body language.

We get over 55% of data through nonverbal correspondence. Allan Pease, an Australian body language master, expounded on this. Emulates, motions and other body developments can expose an individual and mention to you what they truly think or feel.

Chapter 13. Myths About Body Language

We are going to deal with some commonly harbored myths about body language and bust them to paint for you the real picture.

Body Language is Mostly Communication

Imagine having to understand what your friends say by muting your ears and simply watching their body language. Of course, you will fail at it! And odds are you will hate it too! If the world were to go mute and deaf all of a sudden and the only available form of communication left was body language. The entire planet would collapse within a matter of a few hours.

You cannot rely solely on body language to understand a person's behavior. Body language is only one form of communication and not even a major one. It is assistance to words and not the main player in the game. You cannot trust body language to help you understand what a person is saying without listening to their words. However, you can get a good or bad impression of their approach. It is a combination of communication skills that allows you to make your own decision about the honesty or reliability of the person with whom you are communicating. If you do give a clear message that you are honest by keeping your shoulders straight and looking your fellow speaker in the eye, you give a better impression than those who look downward and appear not to be taking much notice of what is being said.

Liars Avoid Eye Contact

One of the oldest myths about body language is that criminals try to avoid eye contact. Imagine if our prison system was based on the mere theory that everyone who avoided eye contact was a criminal. More than half of us would be rotting behind bars if that were the case. It is a fundamental principle of criminology and body language combined that those who lie find it hard to establish eye contact. However, an entire cult of criminals is so hardened and brazen that they have understood the art of lying and have no shame from marinating in eye contact.

They could lie about what they ate in the morning and still look straight into your eyes. Not just that, some of the notorious ones could lie to the extent of misleading and manipulating you to overthink and stress yourself out. The myth that those who have done something wrong would avoid eye contact is a thing of the past.

You can tell from the eyes the level of nervousness of the person with whom you are talking. If their eyes are constantly on the move, this is a good indication that they are not relaxed.

People Who Talk Too Fast Are Liars

Again, another one of those judgmental myths. When people talk fast, it does not mean that they are telling you lies all the way. It is true that some individuals do really talk fast when they are lying—but it does not mean the same the other way around. It depends on the context. Sometimes people talk fast because they are excited about what they are chatting about. Sometimes, that is just the way that they talk.

Maybe they are in a hurry and are trying to drive an important point home. That will force them to speak at a rapid pace and try to convey a message as soon as possible. Not all people who talk too fast are liars. The other thing to bear in mind is that some people are naturally fast talkers. They may have learned that way and may not be that articulate at getting their message across. Often, I have asked people to slow down when they talk in this way, but you will always have people whose level of nervousness comes into their speech flow. It doesn't mean they are lying or trying to hide something. It does mean that they have an inbuilt nervousness about public speaking. Use another body language in conjunction with fast speech to get a more accurate picture of the person.

Crossed Arms Is A Negative Sign

Another very commonly held myth is that crossed arms signify negativity of some kind. Generally, it so happens that in a group discussion, it is considered hostile to cross your arms and not participate in the ongoing discourse. However, crossed arms could mean a lot of other things, each as possible as the next.

Those watching a theatre play and sitting in the front row usually cross their arms in order to establish a sort of barricade between the artists and themselves. In such a case, the front row people feel vulnerable being exposed to the play from such a close angle, and crossing their arms is a form of shielding themselves. If you are in a state where those you are talking to are in a similar situation, such as a lecture room, then crossed arms can simply be doing the same thing. In a situation where manual work is being done, crossed arms may just be the speaker's way of relaxing himself between jobs. However, it still shows defensiveness and a lack of openness to other opinions for a public speaker.

Smiling While Speaking Is an Indicator of Honesty

You must have seen a public speaker employing this method. They tend to smile more when they speak rather than in other situations. Smiling conveys a feeling of security and honesty. However, most of these public speakers are so seasoned that they have mastered the art of smiling to the extent where they naturally smile while addressing a crowd. It is not plastic, but it is not genuine either. The process of smiling while speaking to a crowd has become so ingrained in their system that they cannot help but sprout a smile every time they climb a podium. However, Smiling is not a sure indicator of honesty, as is evident from development-promising politicians who are all smiles right before an election starts.

You will start to identify the difference between a genuine smile and that which would be summed up as "smarm" in order to try and charm an audience into liking a speaker who is not particularly likable. Smarm is the kind of smile that is forced, and when you are talking to others, you can generally recognize one from the other.

Fumbling Is A Sign of Lying

Not necessarily! Some people fumble naturally and have been diagnosed with the medical condition of stuttering from the very beginning. If you remember well, you will recall that we talked about how a lot of factors go into deciding the correct inference about a person's body language behavior.

While we enlisted only five of them, body language myths go on running for three miles at a stretch. It is vital to comprehend that body language is not the ultimate tool for deciphering people's true intentions and emotions. Sometimes, the tool of body language fails miserably. It is not in all circumstances that you can employ body language to give you the most accurate results.

If you assume it to be the ultimate mind reading weapon for you, you are in for a great shock. As it has been mentioned before, consider all the factors possible before applying the tool of body language to decode a person's behavior.

Body language should not be read in a vacuum. There must always be context surrounding a particular body behavior. You are undertaking it all erroneous if you isolate one instance of behavior and decode it. That is a method that will lead you to definite failure. Instead of that, try to observe body language patterns in the surroundings they are created in. Before jumping to conclusions based on one instance, understand that there may be a lot of reasons a person behaves the way they behave. Often times, it is the ongoing situation that makes a person act in a certain way.

With a female, you have to take into account the mood swings the fairer sex is subjected to on account of monthly instances of losing a reasonable amount of blood (menstruation). There could be regional factors too. Most people have their moods on a roller coaster. Mood swings are as common as cupcakes, and you have to acknowledge that people are not constant. Change is the only constant. I may not be the same person today as who I was yesterday. It is, in fact, a good sign that people change for only change could lead to evolution, and without evolution, people would go stagnant, not just evolutionarily but also mentally. Therefore, while trying to decipher a person's true feelings and motives, try to take everything there is into account and then start decoding.

Perhaps the person that you are talking to is nervous in a situation where it would be normal to be nervous. In this case, body language can be forgiven because it's justified.

Any Nonverbal Sign Is A Message

It is one big myth that needs to be busted. Many people assume that any non-verbal sign is a basis of communication. In fact, they believe that 90% of people's communication will come from nonverbal signs. But this is not true. If you accept this to be correct, you might misinterpret things or over-rely on something that does not lead you anywhere. However, you have to practice your own skills of reading your own body language and the body language of people that you have known for a long time. Perhaps you haven't really paid much attention to this, but it's time that you did because it enables you to recognize body language that has a hidden message and that which does not.

Nose Touching Is Indicative of Lying

Nose touching is said to be a universal sign for lying. But this is not true! What if the person scratching the nose has an allergy or is on the verge of sneezing? Maybe the person is really not doing it on purpose. In fact, many people take offense when people touch their nose while speaking. If you think the other person is doing it on purpose, you can simply ask if they have a cold, which will alert them not to do it anymore.

Nose touching can mean various things. It is not a good thing to fix when you are talking to someone, although it is an area of the body that may be giving you problems. You would be better carrying a handkerchief and dealing with the problem, rather than trying to prevent it by touching your nose. Also, this area of the face may have dry skin around it, which may be the reason for the touching. You need to consider each individual case before you assume that someone is lying. In fact, this sign isn't really one of lying at all, and you would be better looking at eye movements and using them as your guide.

Chapter 14. Who Is a Manipulator?

Manipulative humans are the styles of folks that use intellectual and emotional abuse to 1-up you, normally to serve their dreams for energy or manipulate. Disdain the fact that it could be difficult to tell if a person is manipulative while you first meet them. There are numerous developments that manipulative human beings regularly show, which can assist tip you off early to this kind of behavior. It is crucial to appear out for manipulation in a relationship, friendship, or with a family member because you fall prey to a manipulator. It could become challenging to reduce yourself unfastened as soon as you have gotten exceptionally worried in their life. Even though manipulators are, in the end, egocentric, they use numerous schemes and methods to cowl this up. That is why it's so difficult to perceive a manipulative character earlier than it is too late. This list will give you an excellent knowledge of what to look out for in a manipulative man or woman. If you get one or further of these tendencies to your so-referred to as friends, you higher run for the hills. So, without ado, right here are ten bona fide developments of manipulative people you have to appear out for.

Manipulative Human Beings Play the Sufferer

Manipulative human beings are well-known for always playing the victim's function and making themselves out to be extra harmless than they may be. Frequently, they exaggerate or even make up non-public problems in order that others sense sorry for them and sympathize with them. In dating, this trait of a manipulative individual often comes out as dependency or co-dependency. The manipulator may also fake to be vulnerable or weak or want consistent assistance to pull the innocent victim profound into their existence. They do that to attract quality humans to them like a magnet, a good way to exploit later and use them to meet their own egocentric wishes and dreams. With the aid of playing the prey, the manipulator can are searching for out and damage the kindness, mortified conscience, or caring and fostering instinct of the goal. Have you ever had a pal or family member who continuously requested you to lend them money or requested you to buy matters for them, the complete time making you feel guilty for now not having completed so inside the first vicinity? You have probably been coping with a manipulative person. With a bit of luck, you determined your manner out of the entice without too much struggling.

Manipulative Humans Inform Distorted or Half-Truths

Another terrible persona trait that manipulative human beings have is mendacity or distorting reality, so they usually come outright. Fantastic instances of this behavior encompass excuse-making, suppression of important information, underestimations, exaggeration, or hypocrisy. Manipulative humans realize how to bend reality to their advantage. They'll often miss or cover facts to be able to divulge them as being a liar. Manipulators deal with all interactions as though they may ultimately visit trial, and everything they say can be held towards them. As a consequence, they frequently skirt from one place to another the problem or make unclear statements so that when faced, they are able to claim they "never stated that" or that it's miles "no longer precisely what they said."

Manipulative Humans are Passive-Competitive

A similarly demanding character trait of a manipulative man or woman is that they're more frequently than not passive-competitive. A manipulative man or woman might also use this type of behavior to get out of something or to get their manner. They may even do that to make you furious lacking outright deed of something offensive towards you. A family member or friend who frequently forgets something crucial you've got instructed them or overlooks to do something for you that you requested them to perform passive-competitive maybe to control you. It is able to seem innocent, but it's far, in fact, a form of anger, and it isn't healthful for their nicely-being or your sanity.

Manipulative Humans Will Strain You

Manipulative human beings, just like salespeople, will frequently position strain on any other individual in hopes of having you decide before you are truly prepared to. The manipulator believes that you will easily crack and deliver into their desires by using anxiety and managing them. Just like the one's actual-property schemes that stress you to behave fast with the promise of big profits that don't certainly exist, manipulative humans will do something to get you to buy into their sport or advantage a few types of aspect over you. So, be cautious of all and sundry who pressures you to offer a solution earlier than you are equipped, mainly if money is involved.

Manipulative Humans Will Guilt Trip You

A manipulative buddy or family member will often guilt journey you into doing something which you do not want to do, or vice versa, out of something which you do want to do. The underlying cause for this is there, in the long run, selfish personality. Guilt journeys encompass unreasonable blaming from the manipulator, together with concentrated on your soft spot and holding you responsible for their happiness, achievement, or disasters. The manipulator works to goal your vulnerabilities and emotional faintness to coerce you into doing what they want you to do. A manipulative person will frequently make a person they may be in close courting with feel guilty if that character isn't always available. They anticipate absolutely everyone else to assist them in coping with their issues but do nothing in return. Anyone who continually expects you to be the shoulder they cry on, however, who is in no way there for you while you want the same, is most likely a manipulative person.

Manipulative People Provide the Silent Remedy

Have you ever been given the silent remedy from a friend, boyfriend, lady friend, or family member? Probabilities are you had been managing a manipulative person. Manipulative people are bullies. One of the approaches they torment others is with the aid of alienation. Actions like disregarding one man or woman in a collection, now not permitting them to voice their evaluations, or leaving them out are immature strategies used by manipulative adults to claim their dominance. With the aid of showing these behaviors, the manipulative character believes they're coming off as self-confident and powerful. In reality, however, they've low vanity and are extraordinarily self-aware. The handiest way they understand a way to make themselves sense higher is with the aid of hurting others. The next time somebody gives you the silent remedy, don't feel horrific, approximately writing them off completely. It's miles a positive sign of a manipulator and has to be no longer taken lightly.

Manipulative People Do Not Do Anything to Remedy Troubles

Manipulators will, by no means, take the blame for anything. It additionally means that they may in no way make contributions to resolving a hassle in worry that one day they will be held accountable for their movements. A manipulator intends to skate through life while not having to step up and take duty for something. While confronted with something with the aid of a chum or family member, they will both flat out lie and say they by no means did something incorrectly or will make all varieties of justifications for his or her conduct that get them off the hook. You will frequently have many unsettled arguments with a dishonest man or woman, which isn't good. A key sign of this is that a manipulator will regularly quit a controversy or conversation that isn't going their way, without you even realizing it. It's far vital to recognize the way to address warfare well; however, the manipulator cannot do this because they're so centered on themselves and always being inside the proper. Any exact dating may be one in which each human surely needs to assist every other. In case you are coping with someone who can in no way paintings through trouble with you, there is a great hazard that they are now not the proper individual for you.

Manipulative People Choose to Play on Their Home Ground

As we have already set up, the character of a manipulative man or woman is very controlling. A manipulative character will generally insist on assembly or interacting with you in an area in which they feel extra powerful and in control of the situation. It could be their workplace, automobile, domestic, or any other residence where the manipulator senses awareness and ownership. The manipulator, in the long run, does this for two motives. One, they want to hold the upper hand with the aid of being in their consolation sector. And two, they want to weaken you via taking you out of yours. It ought not to be just bodily, both. A manipulator will attempt to take you from your comfort area emotionally and financially as well. Be cautious of everybody who's in no way willing to come out of their comfort zone for you or meet you midway. It's far never an amazing sign.

Manipulative People Rationalize Their Conduct

If ever approached about their manipulative phrases and deeds, a manipulator will make it appear as if it is not a great deal or will shift the responsibility onto someone else, someway making you sense horrific for them usually. However, it's miles, the manipulator who makes a big deal out of things. Until you say something to them approximately it, and then they fireplace every cannon they have got back at you to distract you from the principal subject matter at hand. Manipulators also don't have any empathy for the humans who've helped them and could even pass up to attack those humans, need to experience protecting, or want to cowl up one among their actions or deeds. The manipulative man or woman commonly knows that they have trouble but make it out to look like it's miles the world towards them, instead of the alternative way around. To the manipulative man or woman, not anything they do is ever wrong. Instead, it is always a person else's fault, and there's usually an excuse to rationalize why the manipulative individual said or did what they did.

Manipulative Human Beings Shake Your Confidence

Manipulators regularly cross overboard messing about with different people with the aid of the use of little blow jabs and abuses. Genuine friends ought to sense relaxed poking a laugh at each other harmlessly, but manipulative human beings continually take it a step too some distance. They try this, particularly in groups or social conditions, to undermine others and set up their dominance. Suppose you have a pal that continually leaves you feeling much less than brilliant approximately yourself. In that case, they will be a manipulator, and you ought to cease your friendship with them without delay.

Chapter 15. Dark Psychology and Women

Dark psychology is usually linked with the exploited behavior of people. These types of behavior are perceived very negatively. They often complete successfully for power and resources, and it usually highlights for men, but the samples of women with diversities can also not be neglected. The women's associative behavior that is very antisocial and the trade that hypnotizes and underestimates women's ability to receive and be evil is often taken in very fewer women exploit others. Yet, all of our population don't expect a woman to be threatening. They are often taken very positively, softly, and non-threatening. Even if the women harm, it is minimized, and women are very less responsible for the reactions. Also, they are very less held responsible for the actions, and because of the reason women even have done because the behavior is so unexpected.

It is the reason women think whatever they do, they can always gain sympathy in front of society, which is not necessary but somehow true.

Not everybody knows this dark psychology horror. It would benefit the women, or if they are aware of themselves' darker side, they are afraid of the headed monster. They often don't like to talk about this Complex topic where it is often said that women are the worst Enemies of other women and themselves.

It is very weird to listen to this and talk about it where it is. It is one of the highest growths in society nowadays. Women's empowerment is stronger than any other Era. People are more liberal and more vocal about women's empowerment.

They talk about it more openly. The topic of feminism is so wide and addressed and portrayed in such a beautiful way that everybody comes ahead to give their part and add; however, there is something inside every woman that reacts against their kind goes against their own will.

How it may not look wrong; however, the journey is not so easy, which is a lot of people say that it's a woman who breaks down another woman. Downgrading other Downgrading by Downgrading by life can be hard sometimes anywhere pulling Each Other back. We don't think of it often, and it is a very innocent reality that we are not aware of. It remains in their selves. Town selves' personality is negative, but it is demeaning and taken as negative because that is how we perceive. We know it is a fact that we experience space at every stage of life, but we don't discuss that. We should not discuss that women's empowerment is all about human behavior should be controlled. The women's darkest psychology should also be understood and taken in charge as it is taken in charge of the men. With men, women can also be harmful when it comes to this as compared to men.

Gossiping

There is no boundary set by dark psychology for the people. That's the same with the woman. The backbiting and the gossiping nature are the women's basic nature; it has no boundaries that are defined fun in doing so ever they don't care if anybody sentiments are being hurt.

In contrast, they are being hurt. Women tend to talk so disgustingly about other women they find entertaining; however, it is very shameful because they don't understand how much attention-seeking. It is often taken as a trait of women; however, they don't understand how attention-seeking and how embarrassing it gets, no matter if people are enjoying it, but it still looks really bad. Gossiping might look fine but can ruin another person's reputation because nobody is born a perfectionist, and nobody is born Evil. They are good on the inside; however, if a person is spreading rumors about someone, it can ruin the reputation and affect them similarly, it can of the people themselves.

It is right to be fully solved. They are being offended because if a person is gossiping, for example, if they are gossiping and confronted about their bad habit, they lose control. It does not mean that they stop doing it because it is not bound. They won't do it that openly but they would still do it is because they feel there is nothing wrong with it, and once they are pinpointed for it, they would start to pull away from people instead of abandoning this habit of themselves. They are not that trustworthy to give out secrets to them, and they can also forget someone. When women gossip, they don't know when to stop and what to say, and it can hurt someone genuinely. It gives out negative energy from them. One of the biggest things that women associated with dark psychology give out so many negative Vibes from them that are not our society is used to looking at the home. When they have them too, they always think that they would be that sweet child of people when they are not.

Bad Wording

Usually, females criticize others without giving it a thought, which doesn't take charge of everything. A woman is working to make another woman let down, and they don't feel sorry for it. Older women are bringing down the younger ones; younger ones do it to the old ones. It is an unhealthy practice overall. The funny thing is that whatever they say is so easily digestible to the people that they don't even ask them to shut up then and there. The most common practice of this is in every household between a mother in law and daughter in law relation and even the value educated people tend to do it and the thing that is the part of human nature. However, they don't understand that it can lead to something very aggressive, and still, they don't stop or refrain from doing so.

Glaring

Women are born with the most beautiful eyes, and so are men. The woman is praised for the eyes throughout their life, but it is one of the women's most typical characteristics. They glare others to their soul and give out dirty looks to everybody around them, and it is an incoming threat in general.

We often talk about bringing changes in women's empowerment forever. They can do anything that brings joy to them, and they would judge another woman by clearing at the measure; it would make them feel any better.

But they do it so the other person can feel very uncomfortable and they find betterment in doing so and they would laugh out loud, later on thinking about it. It is a very belittle characteristic of every woman; however, they don't want to help the darker side of the personality, so as a result, they don't stop doing it.

Insecure and Jealous

Women are extremely unconfident as compared to men. The same thing arises when they fear losing anyone they are attached to or something they want badly. They are very fragile, and in that situation, they suffer more emotional jealousy than any other, and it is found in any age of the women regardless of how old they are, and they can do anything to for it. They don't want to lose someone they love no matter what happens. Women react to certain situations, and insecurity and jealousy are quite common in them; however, insecurity becomes an extreme obsession if it is not taken care of.

Comparison and Competition

It sometimes happens that people are very competitive. These are the words in the personality that bring out the best outcome in people, especially women. It gets negative when they demoralize and destabilize and push Each Other down when there is a cold war among them. It just comes with a very competitive nature and then compares them with other women irrespective of any relationship and friendship. It happens quite frequently, which is why we live in a place where there is so much competition going on. Everything is going so digital the competition becomes natural; however, there are two types of competition.

There are one healthy competition and one unhealthy competition among competitions, and women are competing against Each Other. They don't look at the outcomes and what it will bring, which is why it brings out an unhealthy competition between them that brings out the worst in them. It is always necessary to hold the Horses of the hidden and powerful demons, working on the dark side. They are always competing, and they need to know that everything is temporary, and harming anybody while working to get something is not something great to do.

Belittling

The darker side of every human being gives them the feeling that they are superior to others, and they have nothing to do in their life. When that happens to them is that they start thinking that the world revolves around them. They try to belittle others because it is very easy to do so instead of showing gratitude to them, and they do anything to make others wrong. When they are criticized or shown the reality, they won't do anything to make it better. Instead, they always put other people down and want to preserve their superiority over others. They always want to have a high status in front of everybody, so what they think is necessary to do so to look better and everybody, whereas they are just making a fool of themselves, which is quite toxic.

Women cannot usually understand this behavior. People must admit and examine themselves, which is the above dark psychology because it can disturb women's life and people around them because not every woman is like this.

But there are certainly some women whose darker side is more powerful than the other side of their life. Every human being has a dark and light inside. It happens women's approach is how they see different things. Similarly, when feminism has become the town's talk, women should understand that there is a light side of feminism and a dark side of feminism. There is so much more related to the community, especially when their traditional ideas on the concept of femininity. It targets women in a particular way of living in a particular way of acting; however, some women do not agree with the feminism idea and its criteria. They never oppose their beliefs, which are that living like a nice girl is not enough. It may be a pleasant experience to see what it takes to take the whole of the darker side. Still, they need to understand that feminine energy is such a beautiful gift for women to experience by the substrates woman to be only the underside of themselves. They are unaware of how the direct energy works, but they keep falling into a pit hole.

Chapter 16. Techniques of Dark Psychology

Reverse Psychology

A first tactic that a dark persuader can use is reverse psychology. This technique consists of assuming a behavior opposite to the desired one. It is with the expectation that this "prohibition" will arouse curiosity and induce the person to do what is desired.

Some people are known to be like boomerangs. They refuse to go in the direction they are sent to but take the opposite route. It works better when someone else is educated and chooses instinctively rather than thinking about things. They can introduce the intention to do X thing when they suggest the 'do not do X.' When you claim that you will do it, you may wonder whether you will do so.

A dark persuader can use this type of behavior because it is a weakness that the victim has. Take an example of a friend who loves to eat junk food at any opportunity he gets. The dark persuader knows this and will suggest that they eat because it will be good for him, knowing that the friend will choose fast food, anyway.

Reverse Psychology can be used in sales techniques when dealing with a difficult customer.

In this case, the seller can say: "this is a product for rich people. I don't know if it can work for you because it costs a lot of money".

So, the seller is like saying: "I don't want to sell it to you. It's not the right product for you since you can't afford it," just because reverse psychology leads the person to want the product even more.

Masking True Intentions (Door in the Face)

Masking true intentions is another tactic a dark persuader will use to get what they want. A dark persuader will disguise their true intentions from their victims and can use different approaches depending on their victims and the surrounding circumstance. One approach a dark persuader can use is using two requests consecutively because people find it hard to refuse two requests in a row. Take this example; a manipulator wants $500 from their victim. The dark persuader will begin by explaining why they need $1000 while stating what will happen if they cannot come up with that amount. The victim may feel guilt or compassion but will kindly explain to the manipulator that they cannot lend the amount because, quite frankly, it's more than they can manage to give when the persuader lessens the amount to $500, which was what they wanted from the beginning. They will attach the amount with some emotional reason where the victim will be unable to refuse the second request. The dark persuader walks away with the original sum, and the victim is left confused about what took place.

The Blame Game

If the manipulator wants to make you do something against your will, he will have a better chance of getting that behavior by making you feel guilty. Blame is one of the most powerful manipulation techniques known to humankind. Guilt can be used to manipulate people by making them feel inferior to the help and support they have received, or it can also be used to make others feel inadequate for a "condition" they have. Think about all those times you hear people say, "things would be different if I weren't sick." It is one of the most rudimentary ways to make someone feel guilty, but it is very powerful. Besides, you might hear others say things like, "remember when you need my help? Now I need your help." It is a clear attempt to convince someone to follow the manipulator's intentions.

Putting the Other Person Down

Through this technique, we try to make the other person feel less capable than he is. For example, you find every pretext to point out to the victim when makes a mistake, and you do it repeatedly to throw off his self-esteem. A person with low self-esteem is manageable and controllable, therefore manipulable. This way, the manipulator will feel in control of the situation.

When a person tries to manipulate you with this technique, remember that they will attack your identity, telling you phrases like "you are incapable" instead. They will never tell you, "you are behaving like an incapable person." To react to this technique, you have to detach yourself from this psycho-trap. Instead, you need to think that the person is judging your behavior at that moment and not your identity.

Leading Questions

It involves the dark persuader questions that trigger some response from the victim. A persuader may ask a question like, "do you think that this person could be so mean?" This question implies that the person will be bad in one way or another. An example of a non-leading question is, "What do you think about that person?" When we use leading questions, dark persuaders ensure that they use it carefully. Dark persuaders know that when the victim feels like they are being led to trigger a certain response, they will become more resistant to being persuaded.

When the dark persuader feels that the victim has to be aware that he or she is being led, they will immediately change tactics and return to asking the leading questions only when the victim has come down.

Fatigue Inducement

The impact of mental fatigue on perceptual, emotional, and motivational factors are complex. In exhaustion, special effects can be assumed to rely on the operation's essence that causes fatigue. This study investigated the impact of exhaustion on different activities based on working memory demands on brain function and efficiency. The results showed that driving quality was not impaired by exhaustion. The effects of fatigue on novelty therapy depended on the mental requirements for the task that caused fatigue.

Creating an Illusion

Create exaggeratedly high or unrealizable expectations. But presenting and selling them in such a powerful, persuasive, and tempting way for you that you'll end up believing it.

With this technique, it is likely to make the victim see the most beautiful future so that she will be willing to do anything to make it happen, even spend a lot of money. The goal is to make people "daydream" to give them the hope of living their lives to the full.

Commitment and Congruity

Highly skilled and sophisticated manipulators know that building trust capital is essential, especially when building a long-term approach. Think about the most sophisticated conmen you can imagine. These are individuals who take time, often years, building up trust around them through congruent behavior so that others can tumble into their trap.

At first, no one suspects the least bit in this individual as they have earned everyone's trust. As they gain more and more trust, they can use that trust capital to deceive others. It gives them some leeway in case they slip up. Given their track record, they will always have the benefit of the doubt.

This tactic is not common in less-sophisticated manipulators as it requires a great deal of dedication. Impulsive individuals will never be able to pull this off as they focus more on short-term rather than long-term gain. Through this type of tactic, many manipulators can build a name for themselves in their chosen domain. However, they are often exposed. When this occurs, the world is shocked to learn that who they thought was a pillar of their community was actually a manipulator.

One good example of this is a cheating spouse. An individual may cheat on their spouse for years without them noticing what's going on. Then, one day, the manipulator makes a mistake, whatever it is, and they are exposed. The shock that comes to the victim is overwhelming.

The reason why this tactic always backfires is due to the fact that the manipulator doesn't know when to stop. The longer they go without getting caught, the more they think the con will last forever. History has taught us that everyone gets caught eventually.

Reciprocity

It is the classic "quid pro quo," in other words; you scratch my back, I'll scratch yours. However, the victim doesn't know the extent to which they are being manipulated.

A great example of this can be found with informants.

When a manipulator wishes to extract information from someone, they may offer tidbits of information of their own in the hopes of motivating the victim to furnish the information the manipulator is looking for. However, the key to making this tactic work is that the manipulator must give information of little or no value while extracting information that may be profitable.

Manipulators also use this tactic when doing favors. They build up capital and then "call-in" favors. While this may seem like it's perfectly reasonable, it is a manipulation tactic as the manipulator doesn't do favors out of the goodness of their heart. They do it so that they can have people they can rely on in times of need. Alternatively, they can resort to guilt or even blackmail if the other party refuses to cooperate.

Scarcity and Demand

Often, manipulators realize that they have something, or at least have access to something, that people really want. When this occurs, they can manipulate those around them by creating a false illusion of scarcity.

Earlier, we talked about how advertisers generally use phrases such as "limited quantities available" or "while supplies last." These phrases have become so cliché that no one really buys into them anymore.

Yet, manipulators can make this work by creating a sense that there really is a scarcity of a product or service. Some of the more outright, devious ways of pulling this off are by planting fake informants who spread lies. When these lies spread, people may begin to panic and flock to get the products and services in question.

Another way of pulling this off is by spreading rumors on social media. Some people fall for it, and some don't. In the end, the goal is to create enough confusion so that no one is able to tell the difference.

Lastly, manufacturers may go as far as hoarding supplies in order to create an artificial scarcity. It has worked well throughout history. In fact, it's worked so well that it is illegal in most countries. Still, manufacturers can pull this off by controlling the entire supply chain of their products. So, any disruption along that line will cause scarcity, thereby creating panic in people. The manufacturers themselves are not responsible for the scarcity as they are not the ones who technically caused the issue.

Consensus

This tactic consists of setting situations in such a manner that people will agree to them regardless of what it is. Governments do this all the time. For instance, they know that no one will ever agree to a tax hike. Yet, they frame the situation in such a manner that if people wish to continue receiving government benefits, they need to accept the tax hike as there is no other way to fund it. So, people reluctantly accept the tax hike out of fear of losing their benefits.

Chapter 17. Practical Exercises For Mind Control Prevention

The real question here is why, in the first place, would anyone want to control your mind? Some people may not want to check out some of these exercises because they feel like there would be no reason for a person to try to control their minds in the first place, but you must know that there are many reasons people may want to control your mind. Some of the reasons why people would want to control your mind include:

- They want you to get something for them: It may be money, documents, or any other thing. They have chosen you because they know you are the only one that can get it for them. As such, you become their mind control project. There are even stories of people that say that they were robbed one way or the other, but when they checked the security tapes, the people who called the police were the robbers. Sounds strange, right? A professional can get you to rob your own house and plant a bomb in there by yourself, even if you have no bomb training.

- They may want information: This is another reason why someone would want to hypnotize you. You do not necessarily need to have money for someone to need something from you. They may need access codes or maybe the names of people in a place. What they want to do with the information is a total mystery, but the thing is that you might have succeeded in telling them things that you would not normally tell them if you were not hypnotized in the first place.

Exercise 1: Do Not Keep Your Eyes in One Position

People who tend to control the minds of others can be very skilled at times. Some of them would want to use everything they can to get your attention to persuade you and control your mind at all times. When you notice that you are in the presence of someone that wants to control your mind, try as much as possible to keep your eyes in random motion. Do not let your eyes focus on one thing simultaneously, especially if that thing is something they are holding.

There are various ways a person can control your mind, and your eyes are a good gateway for that to happen. You do not want your gateway to be wide open and for you to be defenseless when someone is trying to get into your mind. When someone is trying to control your mind, and you notice, all you have to do is avoid any kind of eye contact with them.

Do not let them think that they can get to you with your eyes because when they do, they will use that technique against you almost every single time. When people like that find your weak point, they tend to exploit it no matter how many times you try to hide it. It is why you mustn't let them know what that is in the first place.

You should not do certain things when trying to avoid eye contact with the person trying to control your mind. These things are said to be very important and should not be taken for granted. Some of these things include:

- Don't let them know: You should never let the person that is trying to control your mind know that you know what he or she is doing and, most importantly. Do not let them know that you are aware of their technique because when they know that you are aware of their technique, they will tend to change it immediately, and they might still be able to get you one way or the other. If you want to be able to get out of that problem, all you have to do is act oblivious.

- Don't get distracted: Getting distracted around a person who is trying to control your mind is the last thing you want to do when it comes to avoiding them. When you want to avoid something like mind control, you need to make sure that you are alert at all times. When you are avoiding the eyes of the people who are trying to control you, you mustn't forget and mistakenly gaze at them again because that might be your downfall. Keep your mind and body alert at all times because the moment you let your guard down, they would not hesitate to take advantage of you.

If you can keep your eyes in constant random motion and at the same time avoid all these pointers, there is a good chance that no one would be able to get into your mind no matter how many times they try. You should know that some professionals would go out of their way to get to you, but if you stick to all that you need to do, you would be one of their biggest challenges. If you play your cards right, you may be able to confuse them to the point that they would have to leave you alone and go for much easier targets. How do you confuse them? When they try to get to you with your eyes, let them get to the point that they think they have almost gotten you and make them know that they are still a long way from penetrating your mind. Once they notice that the closer they are to getting to your mind, the harder it gets; they would get confused because you would become a harder nut to crack.

Exercise 2: Don't Let People Copy Your Body Language

It is probably something that you thought was far from important, but it is. If you are in the presence of a person trying to control your mind and find out the person is sitting in the way you are seated. Even the person is mirroring your movements in any way, keep it in mind that the person is somehow trying to get inside your mind. It is why it is important to mind your surroundings at all times because they could get to you just by mirroring your hand gestures.

You may not notice them doing this because they can be subtle as they possibly can. If you even come in contact with the professionals, there is a huge chance that you will not be able to find out what they are up to until it is too late to go back. It is important to know that you may figure the person out if they are new in the game.

The thing is that professionals are very clean in their game, so clean that you may not know what they are doing until they are done. Still, when it comes to a rookie, you can be able to spot what he or she is doing almost immediately because they are not as clean as the professionals. A professional would mirror your movements and gestures very quietly, meaning that you would never catch them doing it. Still, a rookie, on the other hand, may tend to change his or her gestures immediately. You change yours. That's right; there is a huge giveaway. When you notice something like that happening around you, know that the person you are dealing with is a big-time rookie, and all you have to do is to mess with them and have fun with it. You can change your gestures and movements as often as possible and watch them get confused and break down.

There are certain things that you do not want to be doing. Especially when a person is trying to mirror your movements in any way at all, these things include:

- Never sit in one place: This is probably the last thing you want to do, especially if the person trying to mirror you is right in front of you. When you are in the presence of someone like that, all you have to do is keep moving around. You do not need to move around like a mad person. If not, they would know that you have made them.

Just move around casually like you have no idea what is going on around you, and if you can be in as many places as possible and still make as many gestures as possible, there is a good chance that they are not going to be able to see where you are going. Some of them may get so frustrated. Even decide to get your attention by subtly standing in front of you. It is so that you forget what exactly you are doing. But when they do, you can always change your gestures over and over again. It is to mess with their heads.

- Mind your surroundings: This may be hard for some people because many people find it hard to mind their surroundings, no matter how long you try to teach them. They are more focused on the things happening right in front of them and fail to see the things happening around them.

If you are that kind of person, getting into your mind would be a piece of cake because if you want to notice someone trying to get you, you have to be aware of your surroundings with every chance you get. Do not see something strange on the road or in your house and just let it go like that. Try as much as possible to investigate even if you do not get there by yourself.

- The bottom line to all of this is that if you know what is going on around you, you would be able to tackle and address it before it becomes too late, and when you address it early enough, there is absolutely no way that a person can easily control your mind.

It is imperative to know that these tips would not work for everyone, and you must also know that you would not be able to get the best results out of this if you practice it repeatedly. You need to practice in this context because there are many skilled mind control specialists out there, and you need to be on your game at all times. You do not need to sit down thinking that no one can get into your mind just because you have succeeded in successfully spotting one or two of them coming your way. There is a good chance that you will meet a person that is more than a professional. These mind control specialists do not need to get close to you to know what you are doing and control your mind.

Some of these kinds of people can come to you, and the only thing they have to do is to say a word to you, and that word may be able to trigger some series of events, and before you know it, you are under the control of someone you just met.

Chapter 18. Touch As a Form of Body Language

We engage in touching routinely. We commonly shake hands as greetings or assign to signal shared understanding. Touch, as a form of communication, is called haptics. For children, touch is a crucial aspect of their development. Children that do not get adequate touch have developmental issues. Touch helps babies cope with stress. At infancy, touch is the first sense that an infant responds to.

Functional Touch

At the workplace, touch is among effective means of communication, but it is necessary to keep it professional or casual. For instance, handshakes are often exchanged within a professional environment and can convey a trusting relationship between two people. Pay attention to the nonverbal cues that you are sending next time you shake someone's hand. Overall, one should always convey confidence when shaking another person's hand, but you should avoid being overly-confident. A firm pat on the back communicates praise and encouragement. Remember, people have varied reactions to touch as nonverbal communication. For instance, an innocent touch can make another person feel uncomfortable or frightened.

Touch can become particularly complicated when touch is between a boss and a subordinate. Generally, those in power will utilize touch with subordinates to reinforce the hierarchy of the workplace. It is usually not acceptable for it to occur the other way around. For this reason, you should make sure to be careful even in the instance of using the most trivial of touches and resolve to enhance your communication techniques with your juniors. A standard measure is that it is better to fail but remain on the side of caution. Functional touch includes being physically examined by a doctor and being touch as a form of professional massage.

Social Touch

In the United States, a handshake is the most common way one engages in social touching. Handshakes vary from culture to culture, though. In some countries, kissing one or both cheeks are more common than a handshake. In the same interactions, men will allow a male stranger to touch them on their shoulders and arms, whereas women feel comfortable being touched by a female stranger only on the arms. Men are likely to enjoy a female stranger's touch while women tend to feel uncomfortable with any touch by a male stranger. Equally important, men and women process touch differently, which can create confusing and awkward situations. One should be respectful and cautious. For instance, while you stand close to a stranger on an elevator, it is not acceptable to stand so close to them that you contact him or her.

Friendship Touch

The types of touches allowed between friends vary depending on the context. For instance, women are more receptive to touching female friends compared to their male counterparts. Touch is different depending on the closeness of the family and the sex of the family member. Displays of affection between friends are almost always appreciated and necessary, even if you are not a touchy person. One should be willing to get out of their comfort zone and offer their friend a hug when struggling. Helping others enliven their moods is likely to uplift your moods as well.

Intimacy Touch

In romantic relationships, touches that communicate love play a critical role. For instance, the simplest of touches can convey a critical meaning, such as holding hands or placing your arm around your partner, which communicates that you are together. According to recent communication studies, adults place more emphasis on nonverbal cues than verbal cues when communicating. In the earlier stages of dating, men tend to initiate physical contact in line with societal norms, but in later stages, women initiate contact. Women place more premiums on touch compared to men, and even the smallest of gestures can help calm women. They were upset.

Arousal Touch

Arousing touches are elicited by intense feelings and are only acceptable when mutually agreed upon. Arousal touches are meant to evoke pleasure and involve kissing, hugging, flirtatious touching, and are often intended to suggest sex. One should be careful about their partner's needs. One can greatly improve their communication skills and relationships by considering the nonverbal messages you send via touching behavior.

Additionally, our sense of touch is intended to communicate clearly and quickly. Touch can elicit subconscious communication. For instance, you instantly pull away from your hand when touching something hot even before you consciously process. In this manner, touch constitutes one of the quickest ways to communicate. Touch, as a form of nonverbal communication, is an instinctive form of communication. In detail, touch conveys information instantly and causes a guttural reaction. Completely withholding touch will communicate the wrong messages without your realization.

Ways of Improving Touch in Appropriate Contexts

Pat Someone on the Back When You Grant Them Praise

If your colleague or friend has graduated, earned a promotion, or married, then pat them on the back. Giving a pat suggests that you are happy with the person and are encouraging them. Touch has a therapeutic value that relaxes the mind and the body and helps an individual feel secure and appreciated. At school, you must have felt valued and loved if you were patted on the back.

Initiate Discussions with a Touch to Create Cooperative Relationships

Studies have established that touching a person increases their willingness to cooperate and work with others. They were establishing physical contact with an individual that you wish to initiate a conversation with can help. Sometimes the target person may not realize that you touched them but will register subconsciously and establish a bond.

Extend the Handshakes

Shaking hands shows confidence and simplicity in interacting with others. Touch helps build trust between two people. Make your handshakes firm when shaking hands with people. It is also necessary to remember that some health conditions may make one shy away from shaking hands, and this includes hyperhidrosis, which makes the palms of the person sweat. With sweaty hands, the individual is likely to shun handshakes, and this has little to do with the context of the conversation.

Adjust the Touch-Type Concerning Context

As indicated, touch is highly contextual. For instance, the Japanese do not favor shaking hands, and a person in this environment will avoid shaking hands at all costs. In the American context, shaking hands is encouraged. For this reason, one should adjust their touch-type depending on the contexts. It might be welcome to continuously hold your partner's hands while the same is creepy when talking to a stranger or to a colleague at the workplace.

Another form of touch is tickling, mostly reserved for lovers, parents versus children, and peers. For instance, a mother may tickle her baby, which is a therapeutic touch and is permissible. On the other hand, children of the same age set may tickle each other, which is permissible. However, it is inappropriate to tickle an adult when you are not lovers, or the relationship between you and them is formal.

Touch as a Form of Abuse

Expectedly, there is a thin line between permissible touch and physical abuse. If not, certain one should avoid initiating touch unless fully certain its meaning to the target person. Pushing someone or pinching someone is considered a form of physical abuse. Kicking or striking someone as well as strangling, are forms of physical abuse.

- Touch as a game

In some contexts, a touch is a form of the game, especially teasing. Touch as a form of the game should only happen where the participants are peers and are receptive to it. For instance, your friend or classmate may blindfold your eyes with the palms of their hands from behind. The participants in this tease may touch each other. For instance, the blinded person may try to feel your arms or head to guess the person's identity. In this form of touch, the scope of teaching allowed is large and may be equivalent to that of lovers.

Chapter 19. Facial Expression As a Form of Body

Facial Expressions and Physiognomy

We want to understand by facial expressions all phenomena that we can observe in the face of a human being. By this, we mean both facial features, eye contact, and viewing direction, as well as psychosomatic processes, such as pale. Finally, we also include entire head movements with such. As a nodding, oblique (the latter, depending on the context, of course, the attitude can be assigned).

In general, we are concerned with the evaluation of congruence signals. As long as the facial expressions match the verbal utterances, we usually do not take them very well. When the incongruity is strong, it attracts even the most inexperienced. But the experienced can take note of a variety of facial expressions to perceive even slight disturbances or incipient incongruence (or, of course, first signs of relief, approval, etc.). Often only a barely perceptible grin indicates that someone is making a joke. Or it may be that a (questioning) raised eyebrow is the only indication of contradiction when the other one says, "Yes, I understand what you mean."

At this point in the seminar, the question often arises of how far one can manipulate his non-linguistic signals to what extent z.

For example, it would be possible not to let it be noted whether one grasps or approves of something?

Answer: Of course, anyone can learn to influence his body language to a degree.

However, it is particularly difficult to get the facial muscles under control. So, you can often observe that someone looks outwardly calm because he has learned to control his hands (for example, by intertwining his fingers to prevent him from playing around nervously). Nevertheless, an inner restlessness (if any) will express itself, and most likely, in the face. Why is the manipulation of our facial muscles so difficult? The word "manipulate" includes the word manus (lat., The hand). However, to be able to handle something skillfully, you have to know it well. We do not know enough about our facial muscles to get a grip on them. In general, we do not know how we look or how we affect others. Try it (right now) yourself! Check your facial expression.

A real experiment on this would look like this: You get a small pocket mirror, which you always have at hand shortly. Now and then, you will try first to feel your expression and then immediately see it in the mirror. Ask yourself before and while you look in each case: "How do I look now? How do I now seem to others?" (Or how would I act on others now?)

You will experience very exciting surprises, although they are not positively fascinating for everyone. Some people are horrified when they realize how often they have a discontented, disgruntled look around their mouths and eyes that they did not even realize! However, the less you know about his facial expression, the less you really know him, the less you can, of course, also manipulate him. That is, you have it.

A second mini-experiment that you can do immediately confirms this. After reading the instruction, close your eyes briefly and try to relax your face, especially the lips and chin, as much as possible. Observe and feel consciously what it feels like.

Stop.

Now three questions:

- Have you achieved relaxation?

- Have you got a feel for feeling your facial muscles?

- Were your lips laid together loosely?

If you answered yes to the last question, then you have confirmed what FELDENKRAIS (29) means when he says:

"How is it that such an important part of the body as the lower jaw is constantly held up? Muscles that work while we are awake, without even the slightest sensation that we are doing something to hold the jaw up?

To drop it, you even have to learn how to apply the muscle inhibitor. Suppose one tries to relax his lower jaw so much that he falls through his own weight and opens his mouth completely, so you will wonder how difficult that is. If it finally succeeds, one will notice changes in the facial expression and in the eyes. It will probably also be noticed later that one usually presses his lower jaw upwards or keeps his mouth firmly closed. "

Did the little experiments teach you a little about how little you normally know about your facial muscles? Every actor who deals (or mainly) with pantomime knows the difficulties associated with the conscious creation of a desired facial expression.

Knowing the difficulties of manipulating one's facial expressions is essential if we want to control our facial expressions. With too much control, if they succeed, resulting in a robotic, non-living expression! But Information is also essential if we want to interpret the signals of others. Since the other person is just as unaware of his facial expressions, one can rely on the facial expressions in general quite well.

By the way, the study of the facial expression is divided into two areas, the facial expression itself and the physiognomy. Under the latter, one understands not the momentary, ever-changing expression but the facial features that a person has in general. I call that the "facial expression." If a person often expresses displeasure by squeezing his lips and lowering his mouth's corners, it does not surprise him if he has so-called mis-wrinkles after years. These are deeply scored "lines" that run down from the corners of the mouth. Anyone who looks at the young SCHOPENHAUER's face and then compares it with the old image can clearly see this (see also: "the compressed mouth").

The physiognomy also includes an interpretation of the facial or nasal form. Although the separation from the phrenology, which GALL (94) founded, is not clear. We will not practice physiognomy or phrenology.

Nevertheless, we cannot help but, for example, to register deep scored wrinkles when we consciously perceive. But even such a signal alone has no significance. To be sure, the wrinkling itself is unmistakable, so that we know that this man must often have his lips pinched and the corners of his mouth lowered, but we do not know why this happened. Of course, it may be that this human being is a "Griesgram" who does not like anything. But it may just as well be that this person has suffered a serious illness or a hard fate. Think of persons who have lost a loved one, to people who have spent years in concentration camps or to those who have been tortured (as is commonplace in certain parts of the world today), etc.

It has become customary to assume the following subdivision:

- Forehead area (including the eyebrows)

- Midface, i.e., eye, nose, and cheek area (for most authors, including the upper lip)

- Mouth (or lower lip) and chin area

The Forehead Area

It is believed that the forehead, with its wrinkles and eyebrows, provides information about processes of thinking and analyzing. Although this opinion seems to be a remnant of GALL's phrenology (94), I still hold the forehead's statements applicable. Nevertheless, of course, there is the demand for caution on the part of "scientificity" of such interpretations.

The Midface

The eye, nose, and cheek area are also referred to as the sense of sight. Most authors include the upper lip because they make more nuanced detail statements than we do. We usually only speak of the Lips or from the mouth, so that it is not so important in our frame, where you want to draw the border exactly.

The sense of sight is said to give us clues about taking on the outside world. It is because the eyes are the "window to the world." But they are rightly called the "window to the soul." So that we see that information from the inner life can also be seen in this area. It should also be borne in mind that the mouth also plays a key role in environmental uptake processes.

The Mouth and Chin Area

The mouth has developed from the Ur-Maw, which already has a very simple organism. It represents the relationship to the environment, in that the organism absorbs as well as eliminates it. It is easy for small children to see that they put everything in their mouths to grasp it. Therefore, it is not surprising that the mouth plays an essential role, both when it comes to not "let in" information from the environment and when one does not want to or is not allowed to express.

Next, assign the chin part (including the lower lip) the emotional and intuitive life, and, especially the chin, the assertiveness. A person who is about to assert himself vigorously will push his chin as a mimic signal. (While the assessment of the chin shape regarding the character traits of assertiveness belongs to the field of phrenology.)

And now, let's look at the interpretations of the three facial area turns.

Since we do not want to analyze the forehead's shape, we are concerned with the mimic expressions of forehead wrinkles, horizontal and vertical. Usually, horizontal wrinkles are accompanied by a lifting of the eyebrows. But there is also a barely noticeable lifting of one or both brows, which does not wrinkle.

Horizontal Forehead Wrinkles

As a rule of thumb, we can say that the horizontal forehead wrinkles indicate that the attention has been drawn heavily. However, this strong attention can have very different occasions. For example, Zeddies (94) calls the following:

- Fright.

- Anxiety.
- Obtuseness.
- Astonish.
- Amazement.
- Confusion.
- Surprise.

Again, it becomes clear that individual signals (usually) must be seen in association with others. It also applies within a category, such as facial expressions. For the forehead, wrinkles are automatically associated with the face's other muscle movements, which open eyes (or an open mouth) can lead to. Such a combination provides, for example, the following:

Horizontal wrinkles and open eyes. According to ZEDDIES (94), the two mean signals interpreted together: "The mental attitude lies in a waiting, attentive attitude to any circumstances that offer themselves to the consciousness."

Another possible combination of two mimic signals would be horizontal forehead wrinkles forming in conjunction with half-closed (= easy squinted) eyes. This combination can be observed if someone goes to great lengths to listen; in the case of the hard of hearing, for example, or in situations in which the volume of the transmitter (including technical sound sources such as a radio) is not sufficient. The vernacular describes this with the expression "the ears are pointed." However, this formulation not only describes "in a figurative sense" but also indicates physiological processes. In fact, when we tip our ears, we actually move our severely stunted ear muscles in a reflex that is pronounced in dogs, cats, and rabbits. An additional gesture and attitude change will often accompany the effort to "play" our "spoons."

Chapter 20. Manipulation In a Relationship

The culture romanticizes deceptive relations so much when talking about the love that it can be hard to recognize them for what they are. We have lots of literature suggesting that genuine relationships are about fixation, that pure love is all-out, and that infatuated people have no boundaries or separate lives.

While many people romanticize the concept of a deceptive relationship, we have to realize that it is not real love. Sometimes it may trigger a dramatic storyline and tension that keeps the reader engaged, but there is no fun living through a deceptive relationship that is romantic.

You may have been warned of manipulating people and the fact that coercion and mistreatment are worrying; the facts are that being in a relationship of control and manipulation that never develops into ill-treatment can also be terrifying and dangerous. Just because somebody does not harm you physically does not mean you cannot feel pain from their actions yet.

Being dominated or put down by a partner can damage our faith, make us feel fearful of relationships in the future, and leave us feeling lost rather than comforted, with various mental and emotional injuries with which we should not be burdened.

You may be familiar with the symptoms of a negative relationship. You might have met a partner, for instance, who required you to wear only certain clothing items or did not want you to visit your friends and family.

This person might want to know where you are going, what you are doing, and why you are just a couple of minutes late. Manipulators are frequently very anxious people, allowing nervous thoughts to pass through their brains and control their actions. We channel their intense fear and anxiety into hallucinations about what you might do if you are not around them. They will think about their worst fears and what you can do to damage them, so they will assume you are doing these things when you are not around.

Such things may spur them to hate you if you are not around. Sometimes it may seem flattering to have someone so concerned about you. You might think, "It is so sweet that they always want to know where I am, and I am safe," but it is not their intention when someone is going to take great steps to control you.

Unfortunately, they are not concerned about your well-being. Therefore, the manipulators are thinking, "I need to make sure I know where this person is at all times, so they do not do something that I do not approve of." Your presence is their assurance that you are not meeting their worst fears about the bad things that you are doing to them when you are not both of you together. In this case, they will not be addressing your needs. The manipulator behaves only to serve the interests of his own.

A manipulator will never tell you that but will only be worried about improving the way they look to you. They are always going to use this technique to make sure you feel guilty. They will make you feel guilty if you do not respond for 20 minutes, instead of admitting that it is acceptable for a person not always to write back immediately. They would view you as if you did something wrong or disrespectful to them because, at the time, you were not around your phone or too busy to answer first.

Marriage should feel better, not confining, scary, or distressing, and having an accomplice will make you happier, not more sorrowful. There will be hard times in life. Your mate may not be understood, and they may not understand you. On the way to making you stronger, these challenges should be pure obstacles. There shouldn't be a healthy relationship that continuously drains you and tears you down, making you feel constantly exhausted.

Signs of a Manipulative Relationship

Most of us have had terrible things happening in our lives—enough terrible things that the prospect of a hero sweeping us off our feet and protecting us from any problems for whatever remains of our lives can sound extremely tempting. For this reason, we are sometimes looking in the wrong places for security, empathy, and care.

Reconsider whether your partner's support thoughts include stopping you from making your own decisions and living your own life. This partner secures you by assuming responsibility for your maxed-out accounts. Or perhaps speaking to a partner you have been struggling with does not pay special attention to you; they are trying to make you have no choice but to put all your faith in them and no one else.

A true partner knows they cannot protect you and what it holds from everyday life—they can just support when you need them. If you run into a money-related issue at some point, a trusted partner can help you pray an overabundance of unopened bills—give help, but do not take control of the situation. They will not take your passwords or insist that only a small amount of money per month be allowed until you have paid off all of your current debt. A right partner is going to offer help yet realize you need to manage your problems.

One typical manipulative relationship is making us feel guilty when we see friends and family members. Suppose we imagine someone trying to cut off their partner from their emotionally supportive network. In that case, we envision something similar to the contemptible husband in a movie made for TV that threatens his better half that she will never talk to her closest friend again. Nevertheless, deceptive spouses can also inconspicuously isolate you from your support network.

A shrewdly manipulative person will not outwardly discourage you from seeing your family because it can be an obvious sign that you should be running in the opposite direction. We will make the coercion more subtle, rather than slowly dragging you out of your life, rather than an outright ban. If your partner can convince you to apologize for an action that you know you have not done wrongly and that you are doing, your manipulative partner will realize that he or she can force you to do whatever they want you to do.

Each time you go out with your buddies, your partner can sulk until you blow off other friends just to save the tension. Perhaps your partner will make negative remarks about your loved ones until you begin to believe that the thoughts they have about these people are valid.

You may even have a hobby or an event you enjoy trying to get your manipulator to stop doing it. They will ensure that you know that your interest is idiotic and will ridicule you until you give it up.

The scrutiny of a controlling partner may not always appear as such. It can be framed reasonably and rationally, implying that your partner is just trying to help you. They might even tell you they are trying to help you.

At school, they will research your decisions. Some of their sentences may include: "Why do you choose to use it for your presentation? You are not thinking about what the boss will think? They are going to question your spending habits and how you are going to buy things with questions like, "Did you have to buy another shirt?" Manipulators are going to spin their words, so it is not clear that the choices you make are wrong, but a seed of doubt and insecurity is being planted.

All partners, however, examine each other periodically. Our loved ones are still supposed to look for us, and sometimes we need others to help us make choices or point out bad habits. Remember, always test this person's true purpose and determine why they had wanted you to change your actions.

Sometimes a manipulator may ask for access to your personal belongings in a relationship, but they will not grant you the same rights. We may know all your secrets, but we rarely trust you.

They are not just less likely to share, and they are not helping you.

This type of behavior demonstrates that the other person dominates. Your partner does not reserve the right to search your emails or texts or asking for your passwords because they say they are concerned that you may be cheating. There is a distinction between having insider facts and having healthy independence from your partner, and when you are in a relationship with someone, you do not have to surrender that.

Every so often, sincere couples healing from a disaster would require the weakened spouse to view each other's messages as a form of transparency. If this is not an agreement you have worked out directly with your partner, it is incorrect.

By emotional influence, coercion is all about influencing the way someone else thinks and acts. Coercion is veiled with emotion, or at least what appears to be a sort of empathy. Most of the time, this is a calculated attempt by the manipulator concerned to relate to the victim.

We must recognize the impact it has had on us to overcome this manipulation completely. If you want a healthy relationship with someone, we must look at all the ways we have been affected by their relationship. It may be the first sign that there is a manipulative relationship if that impact is negative.

Most manipulative people have four standard attributes: they know the weaknesses.

- They use your vulnerabilities against you.

- They persuade you to surrender something of yourself to serve them through their quick plots.

- If a controller triumphs in manipulating you, he will likely repeat the crime until the mistreatment is stopped.

- They are going to have a lot of different reasons for keeping you around and controlling you.

One might just be because a past relationship damages them. We may have confidence issues that have made it difficult for them to be transparent and consider other partners. This situation can make them feel like they need to manipulate you to keep you loyal to them.

Understand Your Rights in a Relationship

It can be difficult to understand how to get out when someone is in the midst of an abusive relationship.

Manipulators are good at creating uncertainty, so they can also avoid blame for trying to control others instantly. Recalling your rights is the way to ensure that you are safe.

These are the things you completely have the right to take away and should never allow another person to take away. If you can remember these consistently, manipulation will be easier to confront as it happens, and it will be easier to recognize when a conversation may be toxic.

Then again, you may give up these rights if you convey vulnerability to other people. Our common, main human rights are the following: you deserve respect from others, especially those you respect.

Your thoughts, feelings, and emotions can be expressed.

You reserve the right to understand your own needs, share them with others, and do what you need to do to meet those needs, as long as you do not take anything away from others.

Chapter 21. How to Know if Someone Is Lying Through Body Language

Psychology of Lying

Almost everybody tells a lie once a day or gets lied to. Lying is a part of being a human being with the motive to protect himself against certain situations or to praise oneself. The reasons for lying are endless. Can you remember the very first time you realized that you were lying, or you were lied to?

There is a series of ideas as to why people lie, ranging from saving the hurting of oneself or something else, or with the motive to achieve personal gains. However, science has a different perspective on why people tell lies and the different types of lies. Nobody likes to be lied to, and it's not surprising to find that most liars do not like to be considered as liars.

You wouldn't have any trouble in believing anybody in a perfect world, but unfortunately, it's not perfect, so you need to be cautious about whom to believe and who not to. There are professionals primarily in law enforcement that are trained to detect liars. You don't have to have access to the polygraph machines so that you can understand who is lying to you and who is not. There are many behavioral clues that you could use to know who is telling the truth and who is not.

Detecting deceit will give you the rare opportunity to choose your associates wisely without having to say a word. The body goes into an immense ball of anxiety when a person lies. The trained eye will be able to detect these small variances that occur. Although words may speak their version of the truth, the body never lies. Deceit is the act of covering up the way you truly feel through seeking control. Often, that control is executed in a sloppy manner, thus leading to dominant cues that signal deceit. Whether it's a large lie or a little white lie, the results of dishonesty come with a variety of consequences.

Essentially, people lie as a subconscious form of protection. They are either hiding their negative behavior or protecting their reputations. Even when used to exaggerate a story, they may be attempting to protect the fact that their life is truly boring. They want others to find them enjoyable. Thus, various lies are told.

In general, lying requires more cognitive effort rather than telling the truth because you must work harder and strain to make your information or statement sound authentic. After you have settled on the path of lying, you must remember all the facts, but how? You already changed all facts. Having presented you with the small background about detecting lies, the following are now the various ways you could identify a person lying to you.

Some Liars Are Always Tense and Nervous

It takes a great deal for a liar to pull together fake points to convince you. However, this is not the case with professional liars. These know how to do it just right. But for those who are not used to telling lies, you will quickly notice that their body language is betraying them. On the other hand, a person who tells the truth looks relaxed and happy as far as the story that she is telling is not a sad or painful one.

Some Talk Unusually Slow

If you have ever observed or listened to somebody telling a true story, you might have realized that his or her speech is normal. However, some liars would tend to take quite long before they can respond so that they have a chance to edit their story. They act as if they are trying to be consistent and avoid negative comments. But for other people, it might be hard to detect when they are telling lies, especially salespersons; this is because they have recited lines they keep on mentioning every day with their numerous encounters with customers. You need to keep check of these factors when you are speaking to a person so that you can analyze them and determine when they are telling a lie.

The Hands of a Liar

When people are gesturing and using their hands while telling their stories, this is often seen as a truth-telling sign. However, if the gesturing comes after telling the story, this is often a sign of lying. The mind is so preoccupied with coming up with a story and realistic details that make sense that the mind is too preoccupied to gesture with their hands at the same time that they are talking. Granted, not all people use their hands when talking, but many people do, and this is a simple tactic that the FBI uses and focuses on determining whether someone is lying.

Breathing

Another good indicator if someone is lying is if their breathing suddenly changes. If you ask someone a question and their breathing changes while answering, this is a good sign that they are lying. When somebody is lying, their heart rate upsurges and they turn out to be nervous. It makes them breathe quicker and harder.

Too Still

Another good sign that someone is lying is if they are too still. It is normal for us to move around a bit while talking. It could be shifting in our seat or from foot to foot. Glancing around, hand movements, etc. However, when someone is noticeably too still, this can be a sign of deception. People are often aware that their body language can give them away if they are lying. They think that being fidgety and moving around will give them away. Instead, they do the opposite. They focus very hard on remaining very still so as not to seem fidgety. However, this has the opposite effect than what they were thinking.

Gut Instinct

Lastly, but most importantly, follow your gut instinct. It is probably one of the best ways to figure out if someone is lying to you. People are often very distracted when trying to determine if someone is lying because they focus too hard on the little signs that are supposed to tell you if someone is lying. Frequently, just listen to that other person and then ask yourself if you believe them. We instinctually know when something is "off" about someone. Sometimes we can't even accurately explain what it is or why we feel that way, but we know when something is not genuine. It could be the pitch of their voice, their facial expressions, etc.

Watch the Eyelids

If someone closes his or her eyelids for a long time, it means the person is trying to avoid eye contact. If the person blinks more than three times, it is a sign of nervousness and apprehension that you will catch him or her. If someone uses the hands to cover their eyes, this is another sign that they want to 'block-out' the truth.

Pointing of Eyes

Our eyes point at things we find attractive or where our body wants to go. If you are talking to someone lying, the person will continuously look at the door or watch, signaling the desire to cut short the conversation because they are fearful you will catch the lie.

Avoiding Eye Contact

Breaking eye contact is the most basic way to identify a lie. Someone who has complete confidence about what he or she is saying will never avoid eye contact. However, if someone is lying, he or she will avoid eye contact.

Facial Expressions

Observing facial expressions can help you detect a lie. The most common facial expressions observed in a liar are, dilated pupils, the appearance of lines on the forehead, narrowing of the eyebrows, and blinking eyes. Sweat on the forehead and an angry expression are common with these facial expressions.

Dilated Pupils

Pupil dilation indicates tension and concentration. When someone gets worried about exposure, the pupils unconsciously dilate as they think of ways to hide the lie. If you are talking to someone but unsure if the person is honest or not, look at the person's pupils for answers.

Several key facial indicators may tip you off to whether a person is lying to you. Though none of these are necessarily conclusive in and of themselves, learning to notice these indicators will be your ally when determining if someone is less than trustworthy.

Lines on Forehead

Someone lying may have lines on the forehead because of the stress the person has to bear as they seek ways to cover the lie.

Apart from the facial expression, we can also observe many other gestures in a liar.

Clearing of Throat

If someone is lying to you, he or she will probably clear his or her throat more than once as a nervous tendency to distract from the stress of telling a lie.

Backward Head Movement

When someone is telling a lie, the head could possibly move backward. This gesture occurs as the lying person tries to avoid the source of anxiety because people tend to distance themselves from things they dislike.

Hard Swallowing

The throat of someone who is lying may become dry, and additionally, they may become self-conscious of their swallowing and breathing so as not to give away their deception. Therefore, it is common for a person to swallow hard to bring moisture back to avoid clearing their throat. It is common for people trying to hide a lie.

Statement Analysis to Determine Lie

Analyzing someone's lie through his or her statement is the last step in lie detection. Sometimes what people say does not support their body language. It allows you to detect lies. People often stammer or talk at a fast pace as a way of trying to avoid discovery.

For instance, if you suspect your classmate stole your money and ask her about it, you notice darting eyes and nervousness in her tone. Her body language does not support her statement that she did not steal the money. It means she is lying and has stolen it or knows who did.

No matter how good a person is at lying, if the person's body language is not supportive of their statement, that person is lying. To identify a liar, analyze someone's body language and determine if it matches the person's statement. If the two contradict, you may have a liar on your hands.

You now have a complete idea of analyzing your target by studying body language, expressions, and gestures. It is just one way, however, to analyze people. If you wish to analyze people more efficiently, then you can use the information you gathered from your body language observations. It will give you a complete understanding of your target's state of mind, personality, habits, tendencies, thought patterns, and general operation mode.

Chapter 22. How Body Language Improves Your Mindset

Our body language is the way we speak with our outside world—and the more significant part of us don't understand, we are doing it! Body language phenomenally affects the center of who you are as an individual. It impacts our posture and physiological well-being, yet it can likewise change our psychological viewpoint, an impression of the world, and others' perception of us.

How Our Body Imparts

We utilize our body language to communicate our musings, thoughts, and feelings; we synchronize body developments to the words we express. We impart purposefully through activities like shrugging our shoulders or applauding just as through inadvertent correspondence like twisting in on ourselves or guiding our feet an alternate way toward the individual we talked about. Before spoken language was made, our body language was the primary technique for correspondence. Our body is our major method to speak with life!

How Can It Influence Our State of Mind?

Our body language is how we interface with our outside world, yet it is likewise how we associate with ourselves. How would you treat yourself? Do you slouch over when you walk, or do you walk tall and satisfied? It is true to say that you are thankful for each development that your body makes for you?

Most likely not; we regularly underestimate our body; we frequently decide to condemn it. Body language can impact our physical body and posture. However, it can likewise change how we are feeling. Having a great attitude can affect misery and cause us to keep up more elevated levels of confidence and energy when confronted with pressure.

An up and coming field of psychology, known as installed comprehension, asserts that the association between our body and our general surroundings doesn't merely impact us. However, we are personally woven into the way that we think.

Four Different Ways You Can Change Your Body Language

The followings are four ways you can change your body language.

Flip Around That Glare!

Grinning and snickering is infectious! A complete report on smiling found that a grin which draws in the mouth and moves the skin around the eyes can enact the cerebrum examples of positive feelings. So, grin and grin frequently! Regardless of whether you are having an awful day, grin at any rate! It may very well assist you with turning the day around!

Collapsing Your Arms

The intersection of the arms is a resistance system to ensure the heart and lungs. We regularly do it when we feel shaky, anxious, or disturbed. The physical obstruction gives others the feeling that we are cut off and detached from them.

The intersection of the arms is a broad idea to be an antagonistic body posture anyway. A few investigations have indicated that crossing the arms can cause individuals to progressively industrious when they feel like stopping.

If you believe you need a little additional lift to take a stab at making some regular mindset boosting homegrown cures like Hyperiforce. It contains concentrates of the bloom hypericum frequently utilized as a treatment for low mind-set and gentle nervousness.

Force Presenting

One of the significant specialists in the zone of body language is Amy Cuddy. She made members remain in high force stances and low force models for two minutes before sending them into a top weight talk with the condition. She estimated levels of the pressure hormone cortisol and the predominance hormone testosterone. The outcomes demonstrated that those remaining in high force present had expanded testosterone degrees and lower cortisol levels than those in little force presents.

Quit Slumping

It may appear glaringly evident; however, slumping not just influences your spine. It can likewise change your state of mind! Indeed, slumping can prompt back agony and an irregular spine arrangement. Intellectually, it can leave you feeling miserable, lacking vitality, and shut off from others. Sitting and standing up straighter can assist with settling back torment just as lift your life and state of mind.

Changing your posture can be trying for your body from the outset, particularly on the off chance you are accustomed to slumping over for significant periods! You may feel muscle hurts in the neck, back, and bears—don't stress, this will pass!

Meanwhile, I'd suggest utilizing Atrogel, a natural relief from discomfort cure containing new concentrates of arnica blossoms.

Improve Your Posture to Improve Your Temperament!

Body language likely isn't the first sport you'd think to look at when experiencing a low state of mind. However, investigating our body language can reveal to us how we are truly feeling. Our body language has an immediate connection to our temperament, similarly that our mindset influences our posture.

Simple ways you can fix your posture to adjust your state of mind:

- Smile when you are having a terrible day!

- Unfold your arms when you feel anxious and permit yourself to be available to circumstances.

- Turning the palms of your hands forward when you walk will urge the shoulders to unwind back as opposed to moving advances.

- Power present before pressure instigating situations like prospective employee meet-ups.

Body Language Signs When Someone Hides Something from You

Untrustworthiness. It happens in many connections—and a great deal of the time, it accomplishes more mischief than anything. It's once in a while ever astute to keep insider facts from your accomplice in a relationship. You never need to keep your accomplice in obscurity about a lot of things in your lives together. That is simply out and out insolent. It shows that you don't regard your accomplice enough to recognize that they are deserving of reality. You are saying that they aren't sufficient to be determined what's genuine—and that is, in every case, terrible in a relationship. You generally need to confess all to your accomplice, particularly about vital issues encompassing your relationship.

Be that as it may, a considerable deal of us are childish. Here and there, reality can be difficult to stomach. Now and then, a fact can place us in an extreme condition of a burden once it's uncovered. So, a great deal of us will turn to lie just to spare our butts. Your man may be blameworthy of doing as such. He may be keeping you out of the loop about something that he ought to be opening up to you.

What's more, that is hazardous for a relationship. You can't hope to make your link work if you're not being taken care of the entirety of the best possible realities. You generally need to ensure that you know all that is going on to don't wind up getting tricked or bushwhacked by anything.

Men aren't generally the best verbal communicators. You may likely know this at this point. Be that as it may, he consistently communicates through his body language and physical developments. His intuitive may be disclosed to you many things about himself without seeing it in any event. You simply need to willingly volunteer to ensure that you spot out the signs when they present themselves. You need to ensure that you keep steady over things in your relationship.

Getting and Understanding Nonverbal Signals

Lauren murmured. She'd quite recently gotten an email from her chief, Gus, saying that the item proposition she'd been taking a shot at would not have been closed down. It didn't bode well. Seven days prior, she'd been in a gathering with Gus, and he'd appeared to be extremely positive about everything. Of course, he hadn't looked, and he continued watching out of the window at something. In any case, she'd recently put that down to him being occupied. Furthermore, he'd said that "the task will most likely stretch the go-beyond."

On the off chance that Lauren had discovered somewhat progressively about body language, she'd have understood that Gus was attempting to reveal to her that he wasn't "sold" on her thought. He simply wasn't utilizing words.

The Most Effective Method to Read Negative Body Language

Monitoring negative body language in others can permit you to get on implicit issues or awful emotions. Along these lines, in this area, we'll feature some negative nonverbal signs that you should pay individual minds to.

Troublesome Conversations and Defensiveness

Troublesome or tense discussions are an awkward unavoidable truth grinding away. Maybe you've needed to manage an annoying client or expected to converse with somebody about their terrible showing. Or then again, perhaps you've arranged a significant agreement.

In a perfect world, these circumstances would be settled tranquility. Be that as it may, regularly, they are entangled by sentiments of apprehension, stress, preventiveness, or even resentment. However, we may also attempt to shroud them; these feelings regularly appear through in our body language. For instance, on the chance that somebody is showing at least one of the accompanying practices, he will probably be withdrawn, uninvolved, or miserable:

- Arms collapsed before the body.
- Insignificant or tense outward appearance.
- The body got some distance from you.
- Eyes depressed, keeping in touch.
- Keeping away from Unengaged Audiences

At the point when you have to convey an introduction or to work together in a gathering, you need the individuals around you to be 100 percent locked in. Here are some "obvious" signs that individuals might be exhausted or unbiased in what you're stating:

- Sitting drooped, with heads sad.
- Looking at something different, or into space.
- Squirming, picking at garments, or tinkering with pens and telephones.
- She was composing or doodling.

Step by Step Instructions to Project Positive Body Language

When you utilize positive body language, it can add solidarity to the verbal messages or thoughts you need to pass; on and help you abstain from imparting blended or befuddling signs. In this segment, we'll portray some fundamental postures that you can embrace to extend fearlessness and receptiveness.

Establishing a Confident First Connection

These tips can assist you in adjusting your body language so you establish an extraordinary first connection:

- Have an open posture. Be loose; however, don't slump! Sit or stand upstanding and place your hands by your sides. Abstain from remaining with your hands on your hips will cause you to seem more significant, conveying animosity or craving to rule.

- Utilize a firm handshake. However, don't become overly energetic! You don't need it to get unbalanced or, more regrettable, excruciating for the other individual. On the chance that it does, you'll likely seem to be impolite or forceful.

- Keep in touch. Try to maintain eye contact with the other person for a couple of moments, one after another. It will give her that you're right and locked in. Be that as it may, abstain from transforming it into a gazing match!

- Abstain from contacting your face. There's a typical discernment that individuals who contact their appearances while addressing questions are being untrustworthy. While this isn't in every case valid, it's ideal to abstain from tinkering with your hair or contacting your mouth or nose, especially if your point is to seem to be reliable.

Chapter 23. Proxemics

Now, imagine that you are standing in front of someone. You can see that they are crossing their arms with hands hidden behind them, their eyes shifting nervously from you to veer off to the left now and then. They shift their weight from foot to foot and struggle to maintain eye contact. Something about the body language of this person makes you uneasy, but you cannot place it. They keep their distance from you, and every time you approach closer, you notice that they are likely to move away.

Body language is good at giving us feelings that tell us to be on edge, offended, or relaxed, but if you do not know what you are reading, you will struggle to understand why you feel that way. It can be difficult to know what someone intends to not put meaning to what they are doing. You can have a general idea of how you want to respond, but it can be incredibly beneficial

Proxemics refers to the distance between yourself and someone else—it is the usage of space between yourself and the world around you. Naturally, people put varying degrees of space between themselves and others. When you are looking to understand proxemics, the best way to do so is to consider it a judgment of the relationship between yourself and those around you. You can also judge others' relationships based upon the distance they put between each other, both vertical and horizontal.

The Use of Vertical Space

Vertical space is what it sounds like—it is the space relative to your position height-wise. When someone utilizes vertical space, they attempt to make themselves taller or shorter, depending on the context. Those who want to make themselves taller may want to be an authority or otherwise as someone that is deserving of respect and compliance. They may even use this space when they are trying to look at others who are taller than them—they simply tilt their heads back to look down their nose at the taller person to create the same impact.

When you make yourself smaller, you typically want to be seen as less dominant for some reason. You may be attempting to shrink down to speak to a child to be understood truly, for example, or you may be lowering yourself to make yourself seem more submissive. In particular, people will pull their chins inward when they want to be smaller because they will then be required to look up through their eyelashes at the other person, even if the other person is taller.

The default, eye level, is deemed to be the most respectful—it marks you and the other person as equals deserving of the same respect and consideration.

The Use of Horizontal Space

In horizontal space, you are looking at how near or far people are to each other. You will use this when you are picking apart the relationships of others. There are four distances used between each other, ranging from intimate distances to public distance.

- The intimate distance: This refers to being as close as possible to the other person. When you are in this position, you are usually touching without trying or close enough to do so. It is typically for young children and parents, or for lovers that are comfortable being this close to each other. Generally speaking, this zone is only about 18 inches away from you.

- The personal distance: Slightly further away than the intimate distance, the personal distance covers about 18 inches away up to about 5 feet around you. It is what people are talking about when they say that you are invading their personal bubbles. This zone is usually reserved for those you like or feel comfortable with, such as friends and family members or children who are too old to be within the intimate zone. The closer you can get to the center, the closer your relationship with that other person.

- The social distance: This is a bit further out. It is the distance you naturally try to maintain with strangers around you or interacting with someone else you do not know. Typically, this is between about 5 and 10 feet. You will use this when you are out and about unless you have no choice otherwise. When you are forced to encroach on this distance, you will most often make it a point to ignore the other person in an attempt to ignore the fact that they are violating those personal boundaries, such as sitting on the bus.

- The public distance is even further out. It refers to anything beyond 12 feet and is reserved for instances in which you speak out toward a crowd. You want to be loud enough that everyone in the crowd can speak, so you want to ensure that people are a bit further away from you so they can see and hear you easier. It is reserved for lectures in classrooms, for example, or in performances.

Chapter 24. Muscular Core, Posture, and Breathing

The best way to find out is to copy your subject's muscular core state. Just look at how their muscles are arranged and try to set yours the same way. There's a good expression, "to carry oneself," and your goal will be to carry yourself just like them. Your copy doesn't have to be identical, just close enough. Hence, you feel close enough to themselves. Imitate them as close to perfection as your present acting skills allow (to be a good judge of character, a good analyst, you don't have to be a good actor, but it helps—remember Sherlock Holmes and his transformations?) It isn't hard—just contract whatever they have acquired and kept it that way!

Now, as we learned to carry ourselves like our subject of study, we must learn to walk like them and breathe like them, or at least pretend to do it, deep inside.

Much can be learned from a human posture and walk: people with bad eyesight recognize and spot their relatives and friends by their silhouette, their stance, their walk in the crowd of hundreds of people, alone, as easy as a person with keen eyesight would. Can you stand or sit as your subject does and feel as comfortable as they seem? Can you breathe like them, at the same rate, with the same depth, following the same intervals?

Try and practice it alone at first, looking at a video of someone else. Soon you'll be able to perform it mentally, running the process almost entirely in your imagination. As soon as your musculature and posture imprint feels identical to that of your subject, as soon as your collective breath sounds like one, it's time to analyze their non-verbal message.

Are they demonstrating the will to move closer, shorten the distance between you—or are they trying to distance themselves from you? Is their posture open towards you (face, chest, and groin unobstructed by limbs) or closed from you? (Folded arms, crossed knees, etc.) If their posture is closed, don't jump to conclusions: they may position themselves this way merely for comfort, not because they'd like to lock themselves away from you. If your object's posture is closed and is comfortable—they are likely an introvert. With extroverts, expect abrupt changes in posture, quick movements ahead (lean towards the person they're speaking to, or reach for them), meant to shorten the distance between them.

Body language is a nation-specific feature of communication—in some countries, it's hardly used, while in the other two conversing's, people may resemble two windmills. Still, you can generally detect the heat of discussion by the amount and smoothness of gesturing, even when watching the speakers from a distance. The rougher, sharper gestures become, the less controlled they are, the higher the conflict's likeliness.

A conflict is something often provoked by the opposition, or a third party, with intent to unsettle us, upset us, or make us lose our temper and act out. Our goal in this situation will be to retain control of ourselves. It doesn't mean suppressing our anger or bottling our frustration. It means dissolving the heat of emotions in the cold presence of our reason. It means starting with controlled breathing, restrained posture, and slow relaxation of the muscle core, resetting it to absolute calm.

A person in control is not someone gritting their teeth, holding reins back—it's the person showing calm restraint and conscious choice of their words and actions. Remember the monkey and the computer? The last one is the analyst; the first one lives for battle and spots a good fight mile away. There's a good use for this quality, too: your instincts will tell you when the situation is about to heat up a bit too much, so your reason could be there in time to prevent unnecessary drama before it has a chance to happen!

The point is neither of the two parts of one's consciousness must be restrained or removed from the interaction. When the reason is cast aside, no civilized communication is possible: any conversation will quickly derail and devolve into something childish, silly, and virtually useless for any purposes but socializing itself. If the moving part is suppressed, the person starts feeling discomfort.

It is a significant point. It happens to be twofold: whenever you spot manifestations of discomfort in yourself or your object, you will know it happens because the primal part, the emotional aspect, is subdued by reason. It may occur when the person's reason doesn't want to give something away, yet their body—heartbeat, breathing, perspiration—seems eager to betray them. Hence, they try and shut it off using reason, forcing themselves under control for some time, after which their animalistic part will inevitably act out. You must have seen how leaving the room after a difficult meeting. Usually, people will be overly childish and agitated. They even exclaiming loudly, pushing each other. At the same time, others are craving some sort of physical gratification. It is all the backlash of self-control imposed by reason. Then it is lifted.

Hence, to stay comfortable, to remain in full control of oneself—which is something you want to practice to become a good restrained analyst—one must never suppress their inner feelings! It's hard to give advice on how your computer could keep your monkey in check, as this is a personal thing, inherent to your character. There's a huge number of venting and confidence-building techniques out there, and you're free to try them all! Just remember this simple rule: by indulging a specific whim of your animal, you grow it, not reduce it. For instance, aggressive behavior does not deplete aggression. On the contrary, it increases your aggressiveness—the same as being afraid will not deplete your fear.

Still, techniques help you drop the level of aggression and overcome fear, from the essential things like counting to ten, naming objects around you mentally, or drinking a glass of water—down to counseling and transcendental meditation. In this book, we'll merely say the solution is out there, and self-control is essential if you want to stay an involved yet unbiased party.

On the other hand, this is what you want to notice in the behavior of your subject: not their controlled, reasonable actions, but their slips, their subliminal telltales, the small movements, expressions, and changes in posture that happen without the subject noticing. How to interpret this body language? The problem is that it's inherent to a particular culture and varies from one individual to another.

Many sources claim they're able to teach you some kind of universal list of telltales. One that enables you to tell the truth from lies, present you with recipes of telling an act from the real deal. But these sources are at best-generalized information. It is sometimes applied to many people. Enough to make it seem true, but not to be applied to just everyone. The truth is, only your own experience, attentiveness, and insight will help you to read another person's body language, for there are as many body languages as there are different people.

For instance, when someone is trying to touch or hide a part of their face—lips, the nose, an ear—it's typically considered a sign of secretiveness, the telltale of a person lying or trying to hide some information from the listener. In many cases, it's indeed so—and still, be careful not to call someone a liar just because they tend to rub their three-day stubble while they're thinking.

Another popular facial feature to be pointed out as a telltale: a genuine smile would cause crinkles around eyes, while a fake smile normally wouldn't. Yet again, in many cases, it may be true—we often hear about "someone smiling while their eyes remain cold." Then again, the experiments show the "smiling eyes" can be faked more or less quickly, and if you were to encounter a sociopathic person, someone good at mimicry—you'd never catch them faking a smile.

Approach tendencies in your subject's posture may mean aggression—or they could mean affection, and only your judgment may discern between the two. If your issue demonstrates avoidance tendencies—this, yet again, could mean an entire spectrum of emotions: apathy, fear, disgust, mistrust, submission, meekness, and so on.

A good analyst would always view the non-verbal signals of their subject as a part of the bigger picture, applying to them the knowledge of this person as a whole. Even a habit as simple as biting one's fingernails—are you sure I bite mine when I'm nervous? It may happen a person tends to stick their thumb in their mouth while they're thoughtful, relaxed, their attention directed inward—miles from feeling nervous!

Always remember: what you see is only half of the picture. Another half, no less important, is what you hear.

Chapter 25. Hand Gestures and Arm Signals

It is important to read gestures in the context of other aspects of body language, but we will explore ways of reading gestures. We all talk with our hands often. For some people, the gesturing matches their message well. Some people do not deploy hand gestures while others overuse hand gestures. Most hand gestures are universal. A person that does not use hand gestures may be seen as indifferent.

For this reason, the audience may feel that one does not care about what the other is talking about. If your hands are hidden, then the audience will find it difficult to trust you. If one's hands are open and the palms wide enough, the individual communicates that they are honest and sincere.

Furthermore, randomly throwing hands in the air while talking may suggest that one is anxious or panicking. Extreme anger will also make one throw their hands in an uncoordinated manner. For further understanding, take time and watch movie characters quarreling, and you will note that most people being accused of something will randomly throw their hands in the air. It is something that they have little control over because most of the body language happens at the subconscious level of the mind. Randomly throwing hands in the air indicates that one is overwhelmed with emotions or has given up defending their position in the argument and has left the argument to the individual who started it.

Additionally, one may point at an object or a person. Pointing as a gesture helps the focus of the speaker and the audience to the focused area. During your school days, you probably saw your teacher's point in a particular direction without speaking until the talking students had to stop. As such, pointing at specific students drew the entire class's attention to their direction, making them become the center of attention, and they had to do a quick self-evaluation and stop talking. All these illustrate that body language communicates tone and emotions just as verbal communication.

Furthermore, pointing while wafting the index finger indicates a warning. When one points the index finger at someone and wafts it up and down, then you are denoting a stern warning and judgment to the individual. It is the equivalent of saying, "this is the last warning." Your parent or teacher may probably have a point and waft gesture to signal a warning that what you are doing is wrong and that you should stop. You might have observed that the police or the lead actor uses the index finger to warn someone in movie characters. The finger signal singles out the individual and reduces the focus to just one aspect of behavior that the speaker wants the target person to understand.

If one spreads all the fingers and holds them together against those of the opposite hand, it indicates strong personal reflection, such as praying or remembering the departed soul. The same gesture can be used when one is focusing the mind during meditation or yoga. The holding of each of your fingers against their peers. On the other hand, it may also indicate feeling humble and thankful for everything. For instance, followers of the Catholic faith frequently use this gesture when praying. The gesture shows humility and thankfulness.

Sometimes one may tap on the head once or continuously. When one taps on the head using a hand or a finger, it indicates the individual is thinking hard or trying hard to recall something. For instance, when speaking and you try to remember what another person said, you might use this gesture. Children often tap their heads once or continuously using one finger or the entire palm to signal attempts to recall something. The gesture is equivalent to saying, "Come on, what it was?" or "Come on, what was the name again!" It is a prop to recall hard.

Similarly, a fully raised palm with fingers spread may indicate that one should stop. When stopping the vehicle on the roadside, one raises one of their palms high up, and it is taken as a sign to stop. The same is true in the sporting environment where raising one palm high up commonly communicates that the playing should stop. When arguing with your partner, if they raise one of their palms, it signifies the other to stop arguing or stop whatever action they are doing.

If one claps, the palms together may indicate applauding the message or the speaker. When the speaker is done speaking, the audience may clap their hands together to mark the message's appreciation or both the message and the speaker. However, when the hands are spontaneously and violently clapped, then it is a message that the audience should stop because what they are doing is unethical or irritating. At home, one of your parents probably clapped their hands suddenly and violently to make you stop as well as draw attention to their presence, especially where you were playing loudly around the house.

Relatedly, if one interlocks one hand against those of the other hands and folding them. The application of this gesture indicates that one is attentive but unease at the same time. During an interview, meeting, or a class session, the audience is likely to interlock their fingers and fold them. In a way, the interlocking of the fingers is supposed to offer assurance to the affected person that he or she is safe. One is likely also to use this gesture when he or she is mentioned negatively. Think of how you reacted when you were mentioned among noisemakers or workers having challenges following the company's rules. Most probably, you interlocked your fingers and folded them.

Additionally, if one feels shy or uncertain, the individual is also likely to interlock their fingers and raise the interlocked fingers when speaking. The gesture in this context appears to give some prop for the affected individual enabling them to navigate the anxiety. The gesture in this context is not just about communicating the affected person's physiological status but as a coping mechanism of sudden anxiety and discomfort of the individual.

Still on body language and focusing on gesture, if one raises both hands behind the head and interlocks the fingers, it acts as a cushion for the head. The gesture indicates that one is feeling casual, tired, or simply not tasked by the current conversation. The gesture may also suggest that the individual is feeling tired by the discussion or the activity. Think of how you react when feeling exhausted when talking to a friend or after watching a movie. You probably raised both of your hands behind the head and interlocked the fingers to act as a headrest. In most cases, when one invokes this gesture, the individual is likely to let the mind allow other thoughts to escape from the current conversation.

Correspondingly, there is the gesture where one lets one of their palms to brush down their faces. The gesture is used to signal deeper thinking, process new contradictory information, or accept humiliation in front of the audience. The gesture suggests surrender. It indicates yielding to inner thoughts or views from the audience that one may have initially opposed. At one point, the class or your friends cornered a speaker facing the speaker to pause and take a minute to admit that he or she may have overlooked some facts about the issue. Probably, the speaker used this gesture to indicate defeat.

On the other hand, to indicate rejection or strong disagreement. It is with both hands with palms broad are waved in an alternating manner to create the letter X. You probably drew the letter X using both hands to indicate that you disagree\what is being proposed in class. For instance, as a kid or as a student, you probably drew letter X to signal rejection that you will not follow instructions when the teacher sarcastically indicated that you should not follow his instructions. The sign also indicates retreat to your inner world to avoid listening or watching what the speaker wants.

For accentuation, when hands are open with palms down, at that point, one is communicating that he or she is certain almost what they are talking about. In case your palms are confronting each other with the fingers together. At that point, you're communicating that you just possess the skill around what you're talking about almost. At that point, there's the approximation gesture performed by holding the hand horizontally with palm down and with fingers forward. After that, tilting the hand to the correct and the cleared out. The guess signal shows that an explanation is to be taken a near appraise of the truth.

Equally important, the gesture with a gentle rocking from left to right means that it is not so good or not so bad. The same gesture indicates that an event is equally likely to end in one of the two ways suggesting that it can go either way. The gesture can signal the other person when a match is going, and the friends are watching in the house, and they do not want to wake up the child through loud talking.

Similarly, the beckoning sign has the index finger sticking out of the clenched fist and palm facing the gesture. Then the speaker's finger moves repeatedly towards the gesturer as to invite something nearer. The beckoning sign has the general meaning of commanding someone to where you are standing. The beckoning sign is often performed with the four fingers using the entire hand, depending on how far the sign's recipient is. Depending on the circumstance, when performed with the index finger, it can have a sexual connotation.

If one feels that the speaker is not making sense, they are likely to keep their fingers straight and together while holding them upwards with the thumb pointing downwards. Then the fingers and thumb snap together to indicate a talking mouth. The gesture suggests contempt for a person talking for an excessive period about a topic that the gesturer feels is trivial. In Asian cultures, the gesture is used as a reaction to a dry joke. The gesture may also indicate that one is blabbering.

Also, there's the check signal that's caught on by servers around the world to flag that a supper supporter wishes to pay the charge and get out. The signal is showed by touching the record finger and thumb together and signifying a wavy line within the air associated with marking one's title. Drawing a checkmark within the discussion utilizing the fingers communicates that the person needs to pay the charge.

Chapter 26. Different Types of People and How They Fit In the Social Circle

All of us are full of different flaws that make us feel ashamed. We do have strengths that we want to brag about in front of everyone. Some of us prefer to stay natural in their everyday life while others love to take up their favorite persona to get through different hurdles in their lives. Some people like to make their way by deception, lies, and manipulation, while others prefer to face stumbling blocks but refuse to deviate from the right path. Whatever our choice of being a person in our lives is, the goal mustn't be of hiding our weaknesses and dark spots if we have any. We must allow our flaws to be a part of our personality. We should celebrate our flaws. It is what being human is about. When a person takes up a fake persona, he forgets that the people loving him are loving that persona that he has taken up and not that person who is in hiding under the fake personality. The real success is that people start loving us because of what we are and not because of what we are trying to become.

The Joker

The first category is a joker. The foremost feeling on hearing the word joker is of a person who is cracking jokes and laughing his heart out even during sober conversations. Jokers love jokes, costumes, and makeup. Each makeover gives them a new look and personality. They love to hide their real looks and nature from others. Generally, jokers are considered harmless, but things get different if we bring to mind batman's joker. A scary and nutty person comes to mind who is evil personified. That joker is always bent on inflicting the greatest pain on the people surrounding him. Can you think of a person who fulfills the above personality traits? Do you know anyone who laughs too much, always cracks jokes or tries to tease others while laughing it out? Beware! Jokers are masters of disguise.

The Smart One

Smart people can mold themselves according to the situation. They learn or are naturally gifted to adapt to changing circumstances. Smart people always remember to read other people's styles to gain more knowledge about them. They tend to see through the motives behind their acts and their hidden desires to work with them and gain benefits. Smart people are good at conveying their messages in an effective manner and without making the slightest buzz. They know how to express their feelings clearly, which is the most important thing when it comes to building and strengthening a relationship.

Similarly, smart people are very successful in their businesses or jobs. They work hard to learn how to read people, and the rest gets automatically easy for you. You can tell if a person is smart by looking at how they behave with you and other people around him. One important point to note is that smart people are very good at taking care of their interests, even at others' cost.

The Worker

Workers are the people who belong to a specific social class that is known for doing jobs for low pay only to live hand to mouth in their lives. The jobs they do low demand skills and labor and also have low literacy requirements. This category of people also lives off on social welfare programs. Working-class people mostly remain preoccupied with their day-to-day expenditures. They don't have time to take up different personas and disguises. Also, they are not smart enough to get a job done in the easiest way possible. Their brains are generally wired to do it the hard way. These people typically wear their hearts on their sleeves. They are easy to predict and are simple to understand.

The Loyal

These people are hard to find but exist. They are reliable as well as truthful. If a person is loyal to you, he shares affection with you and will not leave you when life gets hard for you. Loyal people think from their hearts and always work to benefit the people who are close to them. Just like the working class, loyal people are easily predictable and trustworthy.

The Strong

Physically strong people generally have a happy temperament. A strong person has higher levels of physical and mental strength. They don't have self-pity; that's why they are confident and good at judging people and dealing with them. Before they judge other people, they try to judge themselves. Besides, they have higher levels of self-restraint. Their nerves are powerful; that's why they are patient. They also are good listeners and observers. Their physical and mental strengths make them very good at reading other people and reaching an educated judgment. They don't hesitate to ask for help when they are in need, and also, they are open to helping others.

Different Types of Personalities

People are driven by their nature when they do this or leave you wondering why they did something that looked unwanted to you. It is perfectly normal if you think you need to understand someone a bit more than you already do. This someone can be a loved one or a person at our workplace. We have to accept the reality that people are not perfect. We are different, and it is this difference and diversity that makes this world a colorful and interesting place to live in. When people stay true to their role, they tend to contribute their bit to this diverse world. Imagine if we were all created in the same way, how the world looks like then. It would be boring.

Take an example of diversity. When a car hits a motorbike in a road accident, many people gather at the site. Most of them are on-lookers who are just investigating what happened. Some mourn the wounds of the injured while some call the ambulance. Only a handful of them step up and help the injured recover their senses. They try to administer to the first aid and take care of them until the ambulance arrives at the site. It is not that those people leap into a house on fire without thinking about their lives. We react differently to different situations. Our fears and desires trigger these reactions. Sometimes they motivate us, while at other times, they just demotivate us.

In analyzing people, you should know the people around you. What they do and how they react to different situations. Knowing their personality types and the fears that guide their behavior can improve how you interact with other people. It helps you read people more efficiently so that your interaction with them becomes smooth and your analysis of people broadens and deepens. Besides, you can track down your personality traits as well as faults. Let's roll on and take a look at different types of people in the world.

The Reformer/Idealist

The Reformer is a perfectionist. They have principles and are conscientious. These kinds of people have specific ideas to follow, and they come down hard on themselves and other people. They just love to keep them at pretty high standards. They are dedicated and responsible besides having perfect self-discipline.

They are usually successful in life because they tend to get lots of things to happen in a short time, and that too in the right way. They are always looking forward to setting themselves on the right path by eliminating their weaknesses. (9 Personality Types—Enneagram Numbers, nod)

The Performer

As the title suggests, these kinds of people will always be setting goals for themselves. They are highly target-oriented individuals, and they believe in doing rather than sitting on the couch and thinking day and night. They are always striving for success. This drive makes them pretty excellent at doing things right. You can find them in a big company, a shop or on the street selling vegetables or fruit. Wherever they are, their eyes are always on the horizon. They have dreams of success, and they are in the world to make them happen. These kinds of people are considered as role-models by many other people.

They have the fears that drive them toward the top. What makes them perfect is their urge to become somebody. The fear of dying as nobody makes them state-conscious. Instead of discouraging others, they respect the opinion of other people. (9 Personality Types—Enneagram Numbers, nod)

The Observer

These kinds of people spend time thinking and are of an introvert type. Their focus always is on gaining knowledge. They also prefer reading their personality instead of reading others. They remain absorbed in themselves and love to play with different types of concepts. They usually despise worldly attractions like big mansions, cars, and social status. They are always busy searching for themselves. They prefer to observe what is happening in their brains. You can see that these people will lock themselves in their rooms for hours as they love to understand how things go. This exclusive behavior allows them to concentrate on what they do. That's why they are usually considered experts on what they do. As they don't have the social skills needed to keep relationships healthy, they get overlooked most of the time.

The Adventurer

These kinds of people are fun-loving people. You will see them engaged in enjoyable pursuits, and also, they are often in an upbeat mood. They thrive on pleasure and adventures, which makes them a positive person. They tend to avoid negativity at all costs, which helps them fight off pessimism and stress well. They are also very optimistic and don't let tough challenges mar their optimism. They are the ones who always find that silver lining in dark clouds. They stick to that silver lining and turn negative situations fast and well. (9 Personality Types—Enneagram Numbers, nod)

Also, they are highly inconsistent. As they are fun-oriented, they remain in a particular work until the fun factor is alive but shoot out of it once they are bored, no matter if the work is complete or not. Completing projects poses a big challenge to them; that's why they struggle to succeed in the practical world.

The Warrior

As the name suggests, these kinds of people love to throw and take the gauntlet. They are strong and have dominating personalities. You can say they are born leaders and are confident. They are real alphas. They hate to depend on other people and also don't like to reveal their weaknesses. Instead, they use their strengths to cover those people who are around them as their family and friends. They are always ready to take charge of any situation, no matter how thundering and dreadful it is. They love to be the masters of their fate, and they also prefer to take control of people and circumstances.

Conclusion

If your mind is reeling from all the information shared so far, brace yourself. You see, this is an exceedingly vast topic. It is an essential topic because communication is one of the essential parts of our lives. How we communicate impacts our relationships, whether private, personal, or professional.

As with anything else, the impact can be positive or negative, so knowing what your body is saying on your behalf is of the utmost importance. The value in this book is not in learning all you need to know about this subject. It is in understanding that there is so much to know and that you can learn it over time by paying attention and putting in some effort.

Imagine that you are a very shy person who has amazing ideas for inventions or songs or movies, or whatever. Now, imagine how hard it would be for a very timid person to get those great ideas across to the right patent attorney, the right musician, or the right producer if they could barely speak above a whisper when they were nervous.

If they finally did get a meeting with their target audience, how would it look if they averted their eyes and crossed their arms over their body the whole time? Do you think they would be taken seriously? What is the possibility that they would win an influential person in a position of influence over under those circumstances?

There is nothing wrong with any personality type, but if you have a timid personality, know what your body language is saying on your behalf. If that is not what you want to convey, you can learn better behaviors that reflect what you want to say.

What of the person who is the opposite? What if you were naturally loud, bordering on boisterous, and the more nervous you became, the louder you seemed to get?

Being aware of how your volume affects others, you might try to tone it down a bit, but those who are naturally boisterous tend to have "big" body language as well.

If you walk into a room and begin to grip and shake hands as if you were arm wrestling, you would naturally start your event with mistrust and wariness as to your motives though you said very little at a modified decibel.

Here is one last word of caution about becoming a student of body language; never to use one cue to determine what a speaker means. Several factors are involved in each person's dynamic, and all must be considered before making an important determination.

Factors that could possibly affect someone's body language might include a physical or mental disability or limitation, a person's culture or background, or even a current health crisis.

Be aware that you can be influenced by body language with or without your consent, and you can influence others by your own body language, whether you are aware of it or even whether or not you mean to.

Body language is a powerful tool. Understand it and that understanding thoughtfully.

Lightning Source UK Ltd.
Milton Keynes UK
UKHW031126181022
410670UK00014BA/456